SpringerBriefs in Applied Sciences and Technology

SpringerBriefs present concise summaries of cutting-edge research and practical applications across a wide spectrum of fields. Featuring compact volumes of 50 to 125 pages, the series covers a range of content from professional to academic.

Typical publications can be:

- A timely report of state-of-the art methods
- An introduction to or a manual for the application of mathematical or computer techniques
- A bridge between new research results, as published in journal articles
- A snapshot of a hot or emerging topic
- An in-depth case study
- A presentation of core concepts that students must understand in order to make independent contributions.

SpringerBriefs are characterized by fast, global electronic dissemination, standard publishing contracts, standardized manuscript preparation and formatting guidelines, and expedited production schedules.

On the one hand, **SpringerBriefs in Applied Sciences and Technology** are devoted to the publication of fundamentals and applications within the different classical engineering disciplines as well as in interdisciplinary fields that recently emerged between these areas. On the other hand, as the boundary separating fundamental research and applied technology is more and more dissolving, this series is particularly open to trans-disciplinary topics between fundamental science and engineering.

Indexed by EI-Compendex, SCOPUS and Springerlink.

Iffa Mohd Arrif · Solehuddin Shuib ·
Amir Radzi Ab Ghani

High-Flexion Knee Prosthesis

Design and Finite Element Analysis (FEA)

Springer

Iffa Mohd Arrif
Faculty of Mechanical Engineering
Universiti Teknologi MARA (UiTM)
Shah Alam, Selangor, Malaysia

Solehuddin Shuib
Faculty of Mechanical Engineering
Universiti Teknologi MARA (UiTM)
Shah Alam, Selangor, Malaysia

Amir Radzi Ab Ghani
Faculty of Mechanical Engineering
Universiti Teknologi MARA (UiTM)
Shah Alam, Selangor, Malaysia

ISSN 2191-530X ISSN 2191-5318 (electronic)
SpringerBriefs in Applied Sciences and Technology
ISBN 978-981-96-9877-6 ISBN 978-981-96-9878-3 (eBook)
https://doi.org/10.1007/978-981-96-9878-3

This Springer imprint is published by the registered company Springer Nature Singapore Pte Ltd.
The registered company address is: 152 Beach Road, #21-01/04 Gateway East, Singapore 189721, Singapore

Preface

Introduction

The issue of the study is to increase the range of motion or the high-flexion of total knee replacement. Since the conventional total knee replacement could not achieve more than 120°, it is concerning because patients have an increasing demand for high-flexion total knee replacement. However, no explanation could adequately explain why most knees do not flex more than 120° after surgery.

Factors affecting kinematics performance and potential revision sources include flexion range of motion and TKR limitation. These features are intimately associated with the shape of the TKR components. In finding the ideal shape of TKR components, no prior research has employed a strict and methodical approach to design optimisation. More studies and findings should be done regarding the issue to allow better design modifications that could increase knee prosthesis range of motion as it is significant in the human body to do daily activities.

The need to address issues related to allowing a greater range of motion to optimise patient outcomes in TKR surgery motivates the development of new designs for high-flexion knee implants. Hence, the design selection for TKR is significant to consider in achieving a better range of motion.

This book is more than just an academic pursuit; it is a personal journey deeply rooted in my family's experience. My motivation to explore high-flexion Total Knee Replacement (TKR) and advance this field was driven by the challenges faced by my mother and mother-in-law. Both endured the pain of deteriorating knees, relying on painkillers, undergoing gel injections, and eventually having TKR surgery for both knees.

Witnessing their struggles and the long road to recovery fuelled my determination to contribute to this area of research. It is my sincere hope that the work presented in this thesis will lead to improvements in TKR design and outcomes, enabling patients who undergo this surgery to regain their mobility and live life as normally as possible.

Purpose

The book is essential for the surgeon to solve the patient's need for a functional knee prosthesis for daily activities. Thereby, it reduces the revision surgeries and not fully functioning knee prostheses demanded by the patients. Therefore, in the future, the success of total knee arthroplasty is predicted to improve by introducing patient-specific implants and AM technologies which help in saving time.

Acknowledgments

We would like to extend our appreciation to Prof. Amran Ahmed Shokri, Orthopaedic Surgeon, USM, and Assoc. Prof. Dr. Mohd Fairudz Miswan, Orthopaedic Surgeon, UiTM, for their invaluable guidance, expertise, and contributions, particularly in providing medical insights into the knee implant. We are also deeply grateful to Dr. Izhar Aziz and 3D Gens Sdn Bhd for their unwavering support and collaboration. The success of this research would not have been possible without their commitment to fostering innovation and bridging the gap between academia and industry.

We would like to express our sincere gratitude to the Research Management Centre (RMC), UiTM (600-RMC/GIP 5/3 (089/2022)), and the Ministry of Higher Education Malaysia (MOHE) (PRGS/1/2018/TK03/UITM/02/3) for their generous funding and support. This work would not have been possible without their assistance.

Our appreciation also goes to all the staff who provided the facilities and assistance during the testing and manufacturing processes. Their cooperation and willingness to assist were fundamental to the success of this study. A special thanks to our colleagues and friends who offered valuable insights, engaging discussions, and helpful suggestions that have enriched this thesis.

Lastly, we sincerely thank our family for their unwavering support and encouragement throughout the preparation of this book.

Outline

This book covers designing, modifying, and analysing knee prosthesis or TKR components. It started with a CT scan, where the images of a male patient's left knee joint were extracted by 3D Gens Sdn Bhd, a company that specialises in 3D printing. Then, the model of knee joint was developed by 3D Gens Sdn Bhd to create 3D model structures. All these components were designed using CAD software. Then, the design modifications process focuses on the tibial insert component to increase the range of motion. All these design modifications went through analysis using FEA software.

Besides that, additive manufacturing is utilised to fabricate the TKR components using 3D printing techniques such as fused deposition modelling (FDM) and selective laser sintering (SLS). It uses FDM to print the tibial insert with polyamide material. Meanwhile, SLS prints the femoral and tibial components with titanium alloy material. The involvement of 3D Gens Sdn Bhd as an industrial collaborator specialising in 3D printing to fabricate TKR components, including femoral, tibial insert, and tibial components.

Designing TKR jigs was a crucial process for conducting compression testing, as it is necessary to achieve at the end. Consequently, the TKR jigs were fabricated by the advanced manufacturing process using stainless steel material. Hence, compression testing evaluated the deformation when the force was applied vertically to the femoral component. Compression testing results were not discussed further as they were used to validate the results of FEA only. The outcomes of FEA were compared with the compression results in terms of deformation.

Conclusion

The exploration of high-flexion Total Knee Replacement (TKR) and the application of Finite Element Analysis (FEA) have not only been academically challenging but also deeply rewarding. Throughout this process, I have grown both as a researcher and as an individual, gaining insights into the intricate balance of biomechanics and engineering. Finally, we dedicate this work to all the patients who inspire innovation in the field of orthopaedic surgery. May this research contribute, even in a small way, to enhancing their quality of life.

Thank you.

Shah Alam, Selangor, Malaysia
June 2025

Prof. Ir. Ts. Dr. Solehuddin Shuib

Competing Interests The authors have no competing interests to declare that are relevant to the content of this manuscript.

About This Book

In an era where the demand for Total Knee Replacement (TKR) surgery is surging due to rising obesity rates and aging populations, the need for innovative solutions has never been more critical. *High-Flexion Knee Prosthesis* delves into the intricate world of knee biomechanics, offering groundbreaking insights into the design and optimisation of TKR implants. Drawing on cutting-edge research and detailed finite element analysis, this book addresses the pressing issue of post-surgical limitations that prevent patients from performing essential daily activities such as squatting and kneeling.

Through meticulous experimentation and design modification, the author presents a novel approach to creating high-flexion TKR designs that significantly reduce deformation and stress under various flexion angles. With a compelling blend of theory and practice, including the development of customised jigs and real-world testing, this book not only advances the field of orthopaedic surgery but also aims to enhance the quality of life for millions of patients worldwide.

Whether you're a medical professional, a researcher, or someone interested in the future of knee replacement technology, *High-Flexion Knee Prosthesis* offers a comprehensive guide to the next generation of TKR solutions, promising a future where patients can regain their mobility and live life to the fullest.

Contents

About the Authors

Iffa Mohd Arrif completed her Foundation in Engineering at Universiti Teknologi MARA (UiTM), Puncak Alam in 2015. She earned her Bachelor of Mechanical Engineering (Hons.) from UiTM Shah Alam in 2019, followed by a Master of Science in Mechanical Engineering from the same institution in 2024. Her master's research focused on computer-aided design (CAD) for developing implant models and the application of ANSYS software for meshing and finite element analysis (FEA).

Prof. Ir. Ts. Dr. Solehuddin Shuib is a professor of biomechanics at the Faculty of Mechanical Engineering, Universiti Teknologi MARA, Malaysia. He holds a B.S. in Mechanical Engineering from the University of Alabama at Birmingham, Alabama, USA, a master's in mechanical engineering from the University of Toledo, Ohio, USA, and a Ph.D. in Mechanical Engineering from Universiti Putra Malaysia, Malaysia. He has authored or co-authored more than 70 refereed journal publications and presented at over 100 technical meetings. His research interests are in biomechanical engineering and medical devices.

Ir. Dr. Amir Radzi Ab Ghani is an associate professor at the Faculty of Mechanical Engineering, Universiti Teknologi MARA, Malaysia. He graduated from the University of Liverpool, 1994 in Mechanical Engineering. In 1995, he obtained his master's from the same university. He completed his Ph.D. at UiTM in 2014, specialising in the impact performance of metallic structures for automotive applications. Ir. Dr. Amir Radzi has supervised many post-graduate students and published more than 50 papers in international journals and conference proceedings. His current interests are structural impact and crashworthiness, and automotive engineering. His knowledge and expertise in structural analysis enables him to be involved in projects of various areas such as electric mobility, oil and gas, sea transportation and biomechanics.

Abbreviations

3D	Three Dimensional
ABS	Acrylonitrile Butadiene Styrene
ACL	Anterior Cruciate Ligament
AM	Additive Manufacturing
ASTM	American Society of Testing Materials
CAD	Computer-Aided Design
CR	Cruciate Retaining
FB	Fixed Bearing
FDM	Fused Deposition Modelling
FEA	Finite Element Analysis
FFF	Fused Filament Fabrication
ISO	International Organization for Standardization
MB	Mobile Bearing
PA	Polyamides
PCL	Posterior Cruciate Ligament
PE	Polyethylene
PLA	Polylactic Acid
PS	Posterior Stabilised
SLS	Selective Laser Sintering
TKA	Total Knee Arthroplasty
TKR	Total Knee Replacement
UK	United Kingdom
USA	United State of America
UTM	Universal Test Machine

Symbols

°	Degree
%	Percentage
E	Young's Modulus
ν	Poisson's ratio

List of Figures

List of Tables

List of Plates

Chapter 1
Introduction

This chapter includes an overview of the overall research. A brief of the research background regarding the study purpose and research gap is introduced. The problem statement, research questions, and research objectives are explicitly identified. Lastly, the limitations and significance of the study are mentioned in detail.

1.1 Research Background

The human knee is the most stressed and heavily loaded joint as it needs to support the entire human body's weight to do active daily movements such as walking, sitting, running, and kneeling [1]. However, injury and pain from the knee joint are uncertain and unpredictable because of accidents, sickness, or ageing. Total knee replacement (TKR) is a surgical procedure replacing the knee joint with artificial material. The replaced artificial knee joint is called a prosthesis. This surgery and post-treatment are necessary for patients to relieve pain and restore knee joint functionality and durability [2]. Figure 1.1 shows a knee prosthesis consisting of a femoral component, tibial insert, and tibial component [3].

Every country has experienced an upsurge in TKR surgeries in the past few years. Based on Fig. 1.2, the statistics of TKR surgeries show an increasing number of cases from 2014 to 2018. It can be observed that Asian countries such as China, South Korea, and Japan have an increment in the usage of knee implants [4–6]. Meanwhile, in the United Kingdom (UK), the bar chart shows a slight rise in TKR surgeries throughout those years [7–9]. As the United States of America (USA) has a large population, the number of TKR surgeries is also more extensive than in other countries [10, 11].

Furthermore, the number of TKR surgeries are still increasing in countries such as Japan, the UK, and the USA in 2021, as shown in Fig. 1.3 [12–14]. Nonetheless, Malaysia is experiencing an upsurge in osteoarthritis prevalence due to age and

I. M. Arrif et al., *High-Flexion Knee Prosthesis*,
SpringerBriefs in Applied Sciences and Technology,
https://doi.org/10.1007/978-981-96-9878-3_1

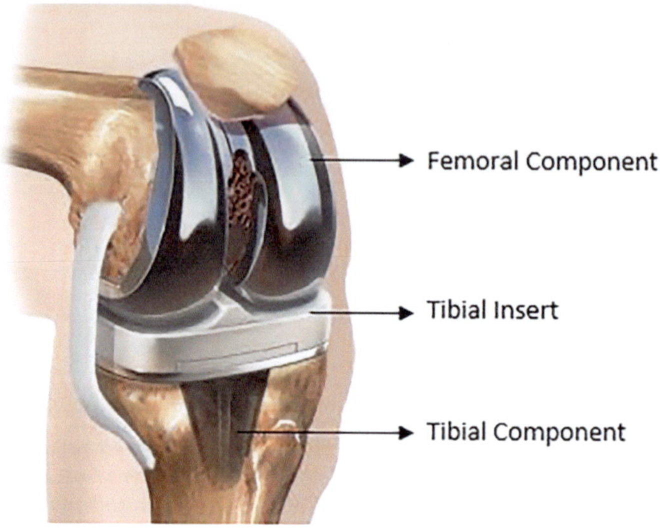

Fig. 1.1 Knee Prosthesis [3]

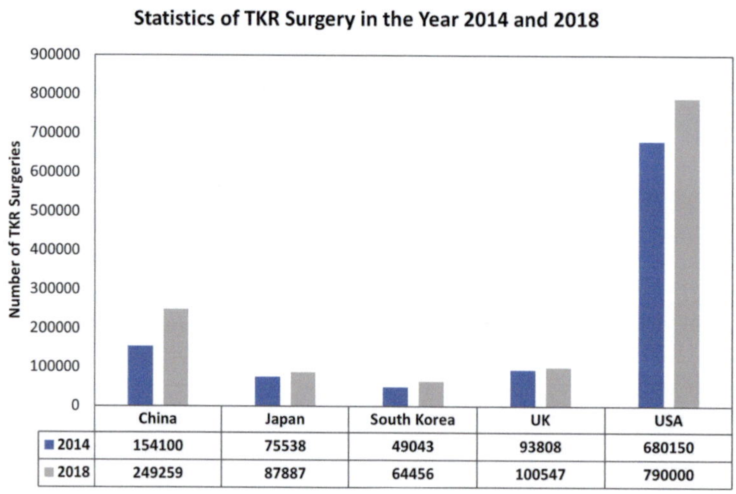

Fig. 1.2 Statistics of TKR surgery in China, Japan, South Korea, the UK, and the USA in the Years 2014 and 2018 [4–11]

obesity, which could lead to TKR surgeries [15]. According to the 2019 Malaysian National Health and Morbidity Survey, 30.4% were overweight, and 19.7% were obese in the country [16]. Hence, the survey concluded that increased age, obesity, and osteoarthritis lead to increased demand for TKR in Malaysia.

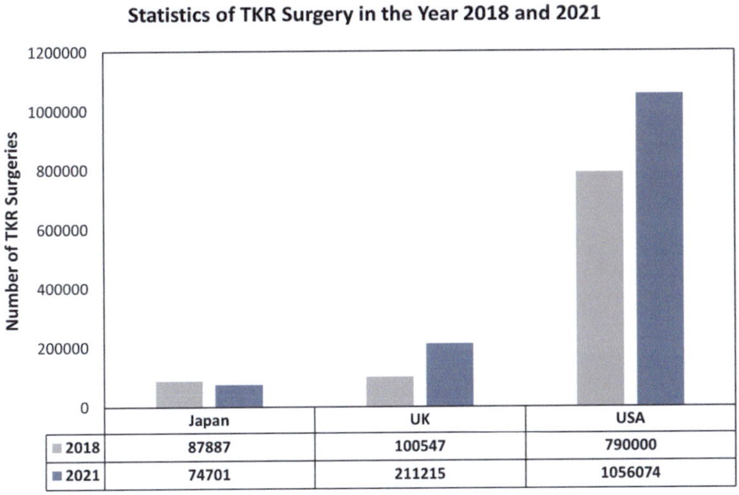

Fig. 1.3 Statistics of TKR surgery in Japan, the United Kingdom (UK), and the United States of America (USA) in the Years 2018 and 2021 [12–14]

Nevertheless, the problem emerges after surgery when patients cannot do daily activities such as squatting, kneeling, and cross-legged sitting because they cannot flex beyond certain angles when performing these tasks [17]. Squatting and kneeling are everyday activities that require normal flexion beyond 90° [18]. However, the extent of the range of motion or high flexion of the knee joint is required in young and active individuals. Besides that, most Asian and Middle Eastern countries must bend their knees up to 165° for high flexion to do their cultural and religious activities [19, 20]. For instance, in Japanese culture, Seiza-sitting is a traditional practice of sitting while attending a tea ceremony, as shown in Fig. 1.4. Individuals in the Seiza-sitting posture need to kneel, placing their knees on the floor and their buttocks resting on their heels. This position requires deep knee joint flexion, often exceeding 165 degrees, to achieve proper alignment and balance.

Meanwhile, several positions require high flexibility for the Muslim religion to perform the prayer, as shown in Fig. 1.5. During Sujood, Muslims kneel and place their forehead, nose, palms, knees, and all toes on the ground, with the hips raised slightly. This position also necessitates significant knee joint flexion, often exceeding 165 degrees, to lower the body towards the ground while maintaining balance and stability.

FEA is an example of numerical computation that uses calculations, models, and simulations to predict the behaviour of an object under different physical conditions. The advantage of using FEA in biomedical engineering is that it helps engineers discover the mistakes in their design prototypes by simulating physical forces. Hence, it reduces the time for trials and prototypes to optimise product components' design [23].

Fig. 1.4 Seiza-sitting from Japanese Culture [21]

Fig. 1.5 Muslim performing prayer [22]

The advantage of additive manufacturing is that the object can be fabricated with complex geometry, which is more complicated than traditional manufacturing methods. 3D printing in medical applications has become common, allowing surgeons and scientists to benefit from additive manufacturing technology [24]. Hence, 3D printing helps orthopaedic surgeons understand and plan their approaches before surgery. Moreover, 3D printing can rapidly produce bone models of various sizes based on a patient's knee. Hence, it reduces the cost of material usage, and the time spent building complex objects in three dimensions [25].

Overall, high-flexion TKR design combined with FEA methods and additive manufacturing (AM) technology allows more opportunity for a fully functional knee prosthesis that enables more flexion to do daily activities. Finite element analysis is an excellent method used to evaluate the range of motion of the TKR. Therefore, the study focuses on modifying the high-flexion TKR design, using finite element analysis to increase the range of motion, and conducting a test on the 3D-printed TKR implants to assess the performance of the innovative TKR design.

1.2 Main Issues

The issue of the study is to increase the range of motion or the high flexion of total knee replacement. Since the conventional total knee replacement could not achieve more than 120°, it is concerning because patients have an increasing demand for high-flexion total knee replacement. However, no explanation could adequately explain why most knees do not flex more than 120° after surgery [17].

Factors affecting kinematics performance and potential revision sources include flexion range of motion and TKR limitation [26]. These features are intimately associated with the shape of the TKR components. In finding the ideal shape of TKR components, no prior research has employed a strict and methodical approach to design optimisation. More studies and findings should be done regarding the issue to allow better design modifications that could increase knee prosthesis range of motion as it is significant in the human body to do daily activities [27].

The need to address issues related to allowing a greater range of motion to optimise patient outcomes in TKR surgery motivates the development of new designs for high-flexion knee implants [28]. Hence, the design selection for TKR is significant to consider in achieving a better range of motion.

1.3 Research Questions

There are a few research questions obtained from the study that aim to be solved. The research questions of this study are as follows:

- How to develop and enhance a design for total knee replacement?
- How to analyse the performance for total knee replacement?
- How to validate the results of FEA?

1.4 Research Objectives

These are several research objectives that must be achieved in the study. The objectives of this study are as follows:

- To design the high-flexion total knee replacement using CAD software.
- To analyse the performance of the total knee replacement in response to deformation, von Mises stress, and contact pressure by finite element analysis (FEA).
- To validate FEA results by comparing them with experimental compression test results of fabricated TKR.

1.5 Scope of Study

This study covers designing, modifying, and analysing a knee prosthesis or TKR components. It started with a CT scan, where the images of a male patient's left knee joint were extracted by 3D Gens Sdn Bhd, a company that specialises in 3D printing. Then, the model of knee joint was developed by 3D Gens Sdn Bhd to create 3D model structures. All these components were designed using CAD software. Then, the design modifications process focused on the tibial insert component to increase the range of motion. All these design modifications went through analysis using FEA software.

Besides that, additive manufacturing is utilised to fabricate the TKR components using 3D printing techniques such as fused deposition modelling (FDM) and selective laser sintering (SLS). It uses FDM to print the tibial insert with polyamide material. Meanwhile, SLS prints the femoral and tibial components with titanium alloy material. The involvement of 3D Gens Sdn Bhd is involved as an industrial collaborator, specialising in 3D printing to fabricate TKR components, including the femoral, tibial insert, and tibial components.

Designing TKR jigs was a crucial process for conducting compression testing, as it is necessary to achieve the last objective. Consequently, the TKR jigs were fabricated by the advanced manufacturing process using stainless steel material. Hence, compression testing evaluated the deformation when the force was applied vertically to the femoral component. Compression testing results were not discussed further as they were used to validate the results of FEA only. The outcomes of FEA were compared with the compression results in terms of deformation.

1.6 Significance of the Study

Firstly, the new design modification of a knee prosthesis was introduced to address the issue of improving knee flexion beyond 120°. This newly modified design has obtained a patent for enhancing knee implant flexion, indirectly contributing to the study. Secondly, the new modified design was produced using additive manufacturing processes, one of the latest technologies for manufacturing products. The advantage of using AM is the complex geometry of the new modified design can be fabricated using a machine directly, which helps save time and minimise labour work. Lastly, the study is essential to solving the patient's need for a functional knee prosthesis for daily activities. Thereby, it reduces the revision surgeries and not fully functioning knee prostheses demanded by the patients. Therefore, in the future, the success of total knee arthroplasty is predicted to improve by introducing patient-specific implant and AM technologies which help in saving time.

References

1. A.R. Balwan, V.D. Shinde, Development of patient specific knee joint implant. Mater. Today Proc. **27**, 288–293 (2019)
2. M. Peto, E. Ramírez-Cedillo, A. Hernández, H.R. Siller, Structural design optimization of knee replacement implants for additive manufacturing. Procedia Manuf. **34**, 574–583 (2019)
3. L.J. Vorvick, D.C. Dugdale, Knee joint replacement prosthesis. Medline Plus (2023). [Online]. Available: https://medlineplus.gov/ency/imagepages/9494.htm. Accessed 26 Jun 2023
4. B. Feng, W. Zhu, Y.Y. Bian, X. Chang, K.Y. Cheng, X.S. Weng, China artificial joint annual data report. Chinese Med. J. **134**(6), 752–753 (2021)
5. Y.B. Kim et al., Trends of total knee arthroplasty according to age structural changes in Korea from 2011 to 2018. Int. J. Environ. Res. Public Health **18**(24) (2021)
6. H. Matsuoka, H. Nanmo, S. Nojiri, M. Nagao, Y. Nishizaki, Projected numbers of knee and hip arthroplasties up to the year 2030 in Japan. J. Orthopaedic Sci. **28**(1), 161–166 (2023)
7. G.S. Matharu, D.J. Culliford, A.W. Blom, A. Judge, Projections for primary hip and knee replacement surgery up to the year 2060: an analysis based on data from The National Joint Registry for England, Wales, Northern Ireland and the Isle of Man. Annals Royal College Surgeons England **104**(6), 443–448 (2022)
8. E. Lenguerrand, M.R. Whitehouse, A.D. Beswick, A.D. Toms, M.L. Porter, A.W. Blom, Description of the rates, trends and surgical burden associated with revision for prosthetic joint infection following primary and revision knee replacements in England and Wales: an analysis of the National Joint Registry for England, Wales, Northern Ireland and the Isle of Man. BMJ Open **7**(7) (2017)
9. J.J. de Gorter, C. Esler, A. Goldberg, R. Beaumont, V. McCormack, Y.B. Shlomo, 12th annual report healthcare quality improvement partnership NJR management team and NJR communications national joint registry. National Joint Registry (2015). [Online]. Available: https://www.hqip.org.uk/wp-content/uploads/2018/02/national-joint-registry-12th-annual-report-2015.pdf. Accessed 7 Apr 2023
10. M. Sloan, A. Premkumar, N.P. Sheth, Projected volume of primary total joint arthroplasty in the U.S., 2014 to 2030. J. Bone Joint Surgery Amer. **100**(17), 1455–1460 (2018)
11. B.R. Levine, B.D. Springer, G.J. Golladay, Highlights of the 2019 American joint replacement registry annual report. Arthroplast Today **6**(4), 998–1000 (2020)
12. Japan Artificial Joint Registration System Secreatariat, THA Registry 2021 Case Statistics. The Japanese Society for Replacement Arthroplasty, 2021. [Online]. Available: https://jsra.info/data/pdf/report-2021.pdf. Accessed 29 Sep 2023
13. R. Brittain et al., NJR Statistical Analysis, Support, and Associated Services National Joint Replacement Registry: 19th Annual Report. National Joint Registry (2022). [Online]. Available: https://reports.njrcentre.org.uk/Portals/0/PDFdownloads/NJR%2019th%20Annual%20Report%202022.pdf. Accessed 29 Sep 2023
14. J. A. Browne, American Joint Replacement Registry (AJRR): 2022 Annual Report. American Academy of Orthopaedic Surgeons (AAOS) (2022). [Online]. Available: https://connect.registryapps.net/hubfs/PDFs%20and%20PPTs/2022%20AJRR%20Annual%20Report.pdf?hsCtaTracking=e22b6617-7eba-4a95-8113-7aa80eb589d1%7C9509b20f-338c-45c0-aeb2-16a46cffd1d2. Accessed 29 Sep 2023

15. Malaysia Health Technology Assessment Section, Clinical Practice Guidelines on Management of Osteoarthritis (Second Edition). Ministry of Health Malaysia (2013). [Online]. Available: https://www.moh.gov.my/moh/attachments/8933.pdf. Accessed 7 Apr 2023
16. Institute for Public Health, National Health and Morbidity Survey 2015 (NHMS 2015) Volume II: Non-Communicable Diseases, Risk Factors & Other Health Problems. Ministry of Health Malaysia (2015). [Online]. Available: https://www.moh.gov.my/moh/resources/nhmsreport2015vol2.pdf. Accessed 7 Apr 2023
17. M. Murphy, S. Journeaux, T. Russell, High-flexion total knee arthroplasty: a systematic review. Int. Orthopaedics **33**(4), 887–893 (2009)
18. T. Nagura, C.O. Dyrby, E.J. Alexander, T.P. Andriacchi, Mechanical loads on the knee joint during deep flexion. J. Orthopaedic Res. **20**(4), 881–886 (2002)
19. M.H. Seman, S. Shuib, M. Afzan, A.A. Shokri, Design and analysis of high flexion femoral component for total knee arthroplasty (TKA). Int. J. Eng. Technol. **7**, 381–385 (2018)
20. J.A. McClelland, J.A. Feller, H.B. Menz, K.E. Webster, Patients with total knee arthroplasty do not use all of their available range of knee flexion during functional activities. Clin. Biomech. **43**, 74–78 (2017)
21. M. Sakamoto, How to sit on a Japanese cushion. All About Japan, Mar. 2019. [Online]. Available: https://allabout-japan.com/en/article/6554/. Accessed 23 Oct 2020
22. Muhammadi, Prayer (Salah)—The Second Pillar of Islam (2021). [Online]. Available: https://muhammadisite.com/prayer-salah-the-second-pillar-of-islam/. Accessed 2 Sep 2023
23. I. Gibson, D.W. Rosen, B. Stucker, Additive manufacturing technologies: rapid prototyping to direct digital manufacturing, 33 Spring Street, New York, NY 10013, USA: Springer US (2010)
24. K.C. Wong, 3D-printed patient-specific applications. Orthopedic Res. Rev. **8**, 57–66 (2016)
25. G.G. Jones, S. Clarke, M. Jaere, J. Cobb, 3D printing and unicompartmental knee arthroplasty. EFORT Open Rev. **3**(5), 248–253 (2018)
26. R. Willing, I.Y. Kim, Design optimization of a total knee replacement for improved constraint and flexion kinematics. J. Biomech. **44**(6), 1014–1020 (2011)
27. C. Batailler, J. Swan, E. Sappey Marinier, E. Servien, S. Lustig, New technologies in knee arthroplasty: current concepts. J. Clin. Med. **10**(1), 1–18 (2021)
28. J.F. Shi, C.J. Wang, F. Berryman, W. Hart, The effect of polyethylene thickness in fixed- and mobile-bearing total knee replacements. Proc. Inst. Mech. Eng. **222**(5), 657–667 (2008)

Chapter 2
Fundamental Design and Material of Knee Prosthesis

The topics connected to this study are thoroughly reviewed in this chapter. The motion and function of the knee are briefly addressed in the beginning. The types of knee prostheses, including cruciate retaining, posterior stabilised, fixed bearing, and mobile bearing designs, are extensively covered. The application of finite element analysis (FEA) and its significance on TKR are also presented. Current fabrication using additive manufacturing technology and testing of TKR are thoroughly examined. Finally, the issues and reviews are summarised.

2.1 Introduction

The knee, one of the largest joints in the body, plays an essential role in daily activities like standing, sitting, walking, running, and jumping. It helps provide better stability to support the entire body weight while standing up and the flexibility required to walk. However, the capability of the knee in high flexion is a crucial element to consider for everyday living activities. Nevertheless, quite a few research have examined the biomechanics of high knee flexion beyond 90° [1].

To begin with, knee joint kinematics involves six degrees of freedom in motion, including rolling, gliding, and rotation. The significant movement of the knee joint is flexion and extension in the sagittal plane with limited rotation. Because of the limitations given by the ligaments, these motions cause a small region of contact between the femoral condyles and the tibial plateaus. Ligaments that connect the femur, tibia, and fibula provide the most support for the knee joint and prevent too much motion [2].

Moreover, the human knee has four ligaments: the anterior cruciate ligament (ACL), posterior cruciate ligament (PCL), medial collateral ligament, and lateral collateral ligament, as shown in Fig. 2.1. The ACL prevents the tibia from turning and sliding forward during agility and jumping. The PCL, which opposes the ACL,

I. M. Arrif et al., *High-Flexion Knee Prosthesis*,
SpringerBriefs in Applied Sciences and Technology,
https://doi.org/10.1007/978-981-96-9878-3_2

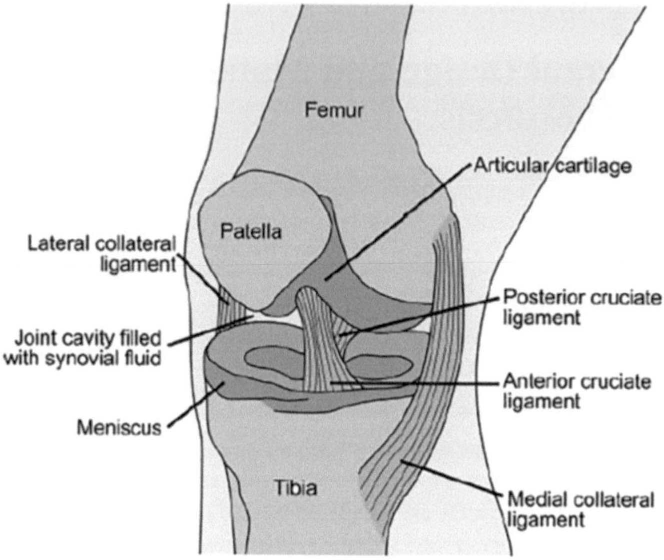

Fig. 2.1 Bony structure and soft tissues at anterior view [3]

inhibits tibia rearward sliding and is situated directly behind the ACL. The medial and lateral collateral ligaments on either side of the knee joint stabilise the knee against slipping from side to side. Therefore, these ligaments work together to establish stability and prohibit excessive movement of the femur and tibia in various directions.

2.2 Total Knee Replacement (TKR)

Total knee replacement (TKR) can be described as replacing a knee joint with knee prostheses by conducting surgical procedures to help ease pain and restore walking ability in patients suffering from severe knee injuries and diseases [4]. Total knee replacement (TKR) or total knee arthroplasty (TKA) is one of the most effective procedures to date, predicted to rise in the future years, with 3.48 million TKR procedures performed by 2030 [5]. The success of the surgery and patient satisfaction with TKR have increased over time because of preoperative assessments, surgical procedures improvement, technological advancements, and orthopaedics research [6].

The discovery of the first condylar knee prostheses with an anatomical approach was in the 1960s by an orthopaedic surgeon, Frank Gunston, inspired by Sir John Charnley's discovery of a metal-on-polyethylene total hip arthroplasty [7]. In 1968, the design of an unhinged prosthesis that replaced the lateral and medial sides of the knee and used independent condylar components was implanted for the first time. The implant, composed of stainless steel in the femoral component alongside plastic tibial, immediately gained favour despite not being entirely condylar [7]. The components involved in the TKR are the femoral component, the plastic insert, the patella, and the tibial component, as shown in Fig. 2.2.

To introduce components in TKR: first and foremost, the femoral component is made of metal that curves around the end of the femur bone. The femur bone is cut to fit the size of the femoral metal component so the kneecap can move smoothly against the femoral metal component as the knee bends and straightens. Second, the tibial component has a stem attached to a flat metal platform inserted into the centre of the tibia bone. A plastic spacer connects to the flat metal platform that acts as a cushion between the femoral metal component and the flat metal platform of the tibial component. Lastly, the dome-shaped component attached to the kneecap is called the patellar component, which is made from polyethylene. In some cases involving total knee replacement surgery, the kneecap or patella does not need to be replaced. Hence, the critical focus of the total knee replacement consists of the femoral, tibial, and plastic insert, known as the tibial insert [9].

The stability and functionality of the knee joint are greatly impacted when any component of a knee implant is missing. It would greatly hinder movement, raise the possibility of additional harm, and make the patient uncomfortable and in pain. The absence of a femoral component would result in a lack of stability and support for the knee joint because of no replacement for the distal end of the femur, leading to significant issues with weight-bearing and movement. The missing tibial component would seriously compromise the knee joint's strength, making weight-bearing

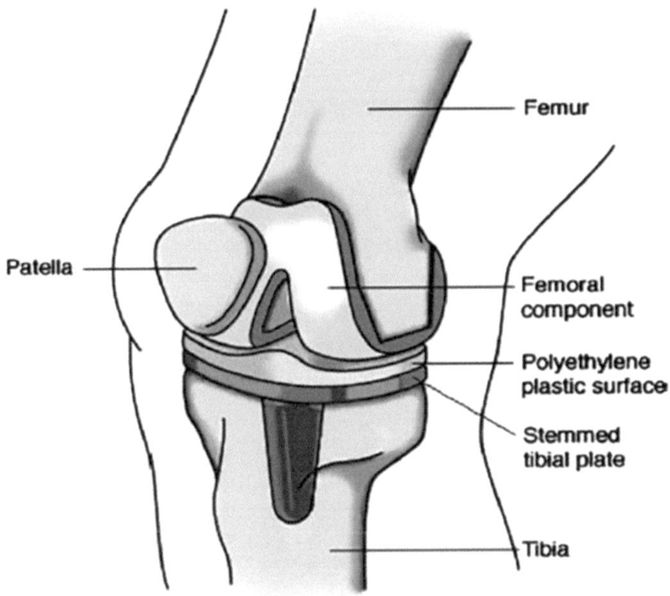

Fig. 2.2 Components of total knee replacement (TKR) [8]

exercises extremely difficult. The absence of a tibial insert would increase friction and wear between the metal components, leading to accelerated deterioration of the artificial joint. Not just that, the angle and position of the plastic insert play a crucial role in determining the knee joint's range of motion and stability. Incorrect placement can result in restricted movement, impingement, and discomfort for the patient. Therefore, proper alignment and positioning of all components are essential to getting the most significant possible outcomes and raising the patient's quality of life after TKR surgery.

2.3 Design of Knee Prosthesis

The current total knee replacement is differentiated into non-constrained, semi-constrained, and constrained, as shown in Fig. 2.3 [10]. Firstly, in a non-constrained design, the stability of the knee joint relies entirely on the ligaments and soft tissues of the body [11]. Secondly, the prosthesis design partially restrains the femoral condyle in a semi-constrained design. Lastly, in a constrained design, a central post between the femur and tibia provides stability in the anterior–posterior and medi-olateral directions. Table 2.1 shows the different types of knee prosthesis designs [12].

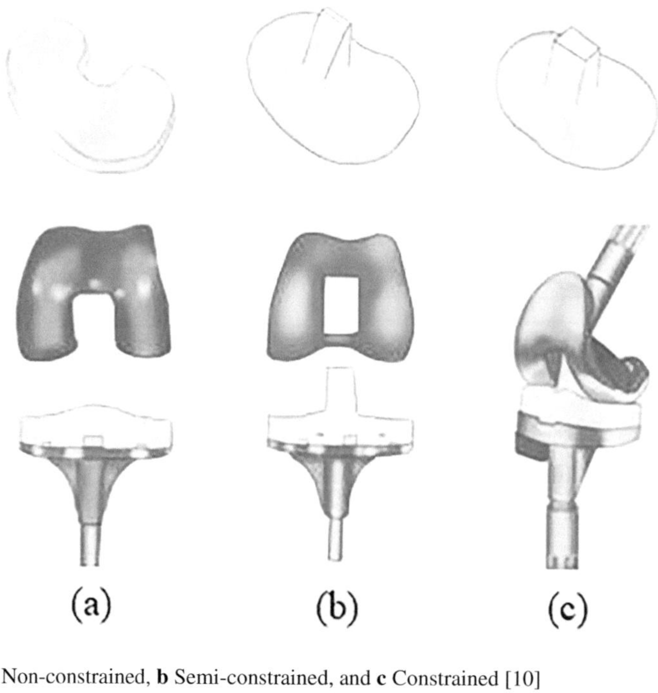

Fig. 2.3 **a** Non-constrained, **b** Semi-constrained, and **c** Constrained [10]

Table 2.1 Types of knee prosthesis designs [12]

Non-constrained	Semi-constrained	Constrained
Preserved the ACL and PCL, as shown in Fig. 2.4	Partially restrains the femoral condyle and sacrifices PCL, as shown in Fig. 2.5	The femoral component is physically linked to the tibial component Does not require ligamentous or soft tissue support
The stability of the knee joint relies entirely on the ligaments and soft tissues of the body	The stability of the knee due to the cam and post mechanism, as shown in Fig. 2.6	A central post between the femur and tibia provides stability to the anterior–posterior and mediolateral
• Bi-cruciate retaining design • Cruciate retaining design	• Posterior stabilised design • Rotating hinged design	• Total stabilised design • Hinged design
• A damaged or torn PCL can cause flexion instability and dislocation	• The more distal femur bone is removed to balance the extension gap • The cam can spin and dislocate over the post with a loose flexion gap	• Most of the bone must be resected

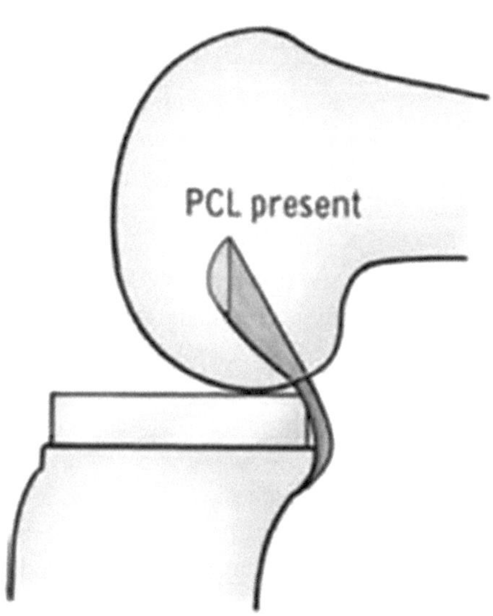

Fig. 2.4 Posterior Cruciate Ligament (PCL) present [13]

The post and cam mechanism can be seen as shown in Fig. 2.6. The design comprises two independent components that glide against one another, but the physical properties of the tibial component prohibit excessive femoral glide. The reason is due to the cam and post mechanism that forces the femoral condyle to move posteriorly during flexion.

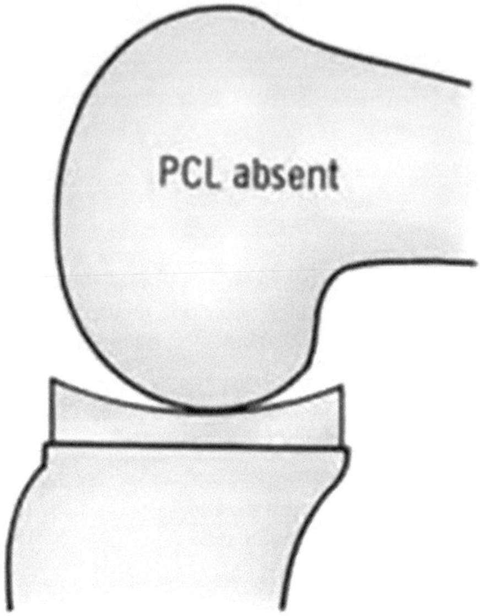

Fig. 2.5 Posterior Cruciate Ligament (PCL) absent [13]

Non-constrained and semi-constrained are suitable for increasing the range of motion of TKR, such as cruciate retaining (CR) knee prosthesis and posterior stabilised (PS) knee prosthesis, respectively. The constrained design is incompatible as it constricted movement for flexion. CR and PS designs are fixed bearing (FB) knees, where the plastic insert is locked to the tibial component. Nevertheless, the difference is CR knee prosthesis preserved PCL instead of removing it as PS knee prosthesis during total knee replacement. On the contrary, the plastic insert is not fixed to the tibial in the mobile bearing (MB) design. However, the difference is that the PCL must be removed during total knee replacement for both PS and MB designs. In other words, MB knee implants are an alternative to FB knee implants, such as CR and PS knee implants, which help reduce wear and improve range of motion (ROM) with a particular emphasis on younger patients.

Furthermore, these types of designs for knee prostheses are introduced to achieve the goal of restoring knee functionality and lessening pain in patients. The cruciate retaining (CR) prosthesis means the posterior cruciate ligament is retained during TKR, and the CR prosthesis design is shown in Fig. 2.7. Some of the advantages and disadvantages of CR prostheses are listed in Table 2.2.

In contrast to CR knee prosthesis, the posterior stabilised (PS) design means the PCL is removed during total knee replacement, and the design for PS knee prosthesis is shown in Fig. 2.8. Some of the advantages and disadvantages of PS prosthesis are listed in Table 2.3.

Between fixed bearing designs, there has been endless discussion about whether CR or PS knee prostheses are preferable for achieving a better range of motion [19]. The reason is that some believe retaining more of the original anatomy inside the human body is vital so the PCL can keep stabilising the knee during flexion

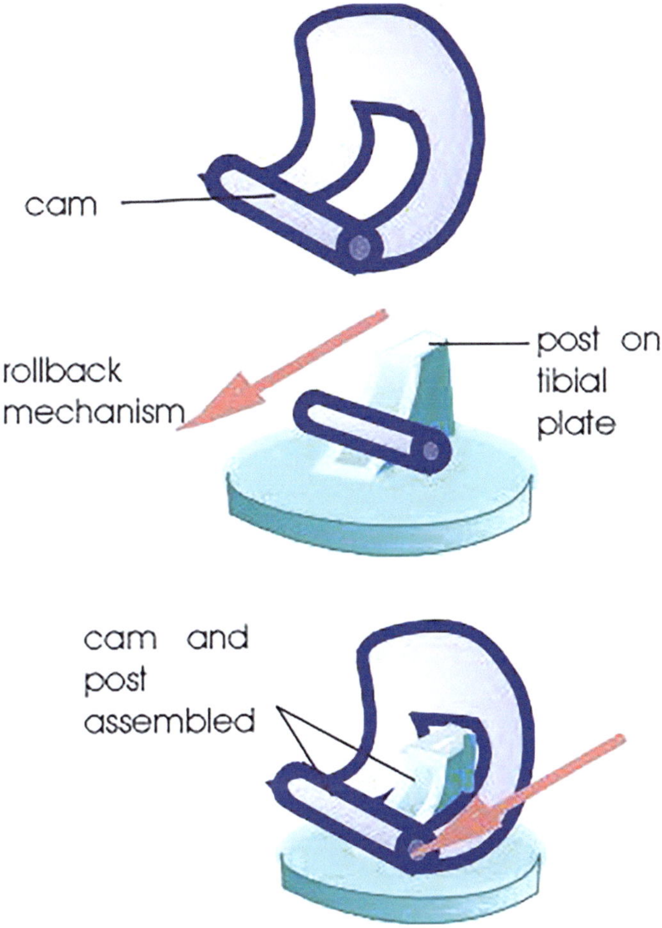

Fig. 2.6 Cam and post mechanism on total knee replacement [14]

[20]. However, most studies have failed to justify which type is better in terms of functional recovery, knee extension or range of motion (ROM), kinematics, and a chance of survival in the long term. The reason is that each type of design for knee prosthesis has its strengths and weaknesses [21, 22].

Nevertheless, there are some issues regarding how surgeons choose the type of prosthesis according to their skill and practice because CR and PS prostheses require different surgical procedures. It has happened to be a complicated process to perform the surgical procedures for CR prostheses [19].

However, the posterior stabilised design uses a tibial post and a femoral cam to replace the posterior cruciate ligament, allowing femoral rollback [23]. In short, femoral rollback can be defined as a posterior translation of the femur component over the tibial component during knee flexion, and in vitro TKR studies show that restored

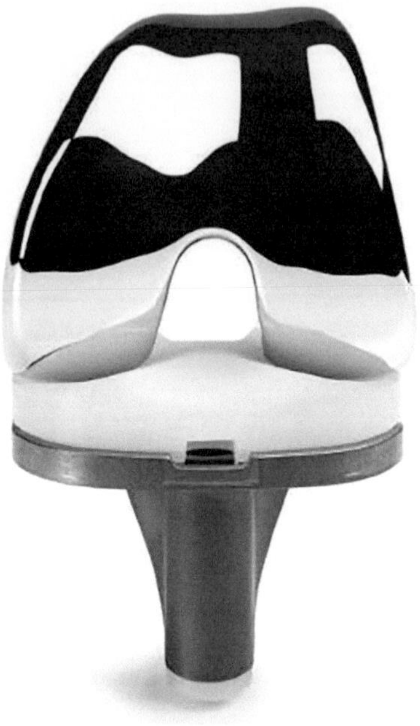

Fig. 2.7 Cruciate retaining (CR) Prosthesis [15]

Table 2.2 Advantages and disadvantages of CR knee prosthesis [16]

Advantages	Disadvantages
• More bone preservation • More normal knee kinematics • Allowing femoral rollback on the tibia during flexion • Offering more prosthesis stabilisation	• Insufficient PCL • Unstable posterolateral • Significant coronal deformity • Arthritis inflammation • A deficient extensor mechanism

physiological femoral rollback helps in reducing quadriceps and patella loads [24, 25]. Therefore, designs that restore the most physiological femoral rollback have longer flexor muscular lever arms, require fewer quadriceps to extend the knee under load, and have a lower patellofemoral compressive force.

2.4 High-Flexion Knee Prosthesis

A high-flexion knee prosthesis design is introduced to increase the range of motion. The knee joint range of motion can be referred to in Fig. 2.9, where 0° indicates a complete extension and 180° is fully bent. The post-surgery range of motion for TKR is generally less than 120°.

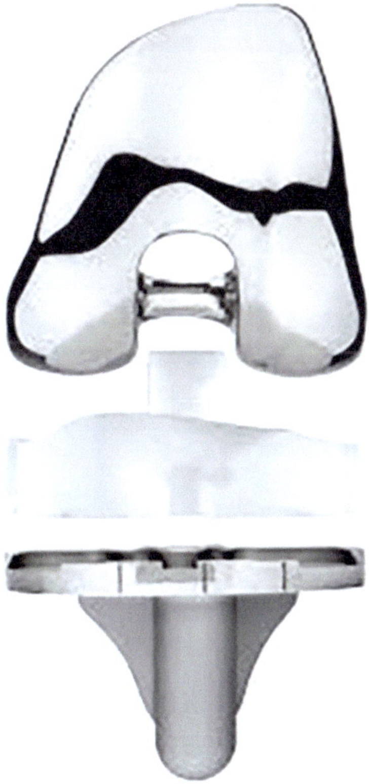

Fig. 2.8 Posterior stabilised (PS) Prosthesis [17]

Table 2.3 Advantages and disadvantages of PS knee prosthesis [18]

Advantages	Disadvantages
• More predictable kinematics • Increased range of motion • Improved ligament balancing • More straightforward surgical procedure	• Tibial post wear • Femoral cam dislocate

A few studies reported that high-flexion TKR might contribute to minimising contact loadings and a greater range of motion than prior TKR. For instance, according to Huang et al., patients with high-flexion knee prostheses experienced average flexion of around 10° more than corresponding patients with conventional PS knee prostheses [27]. It is due to a larger contact surface during deep bending, which distributes contact pressure across a wider area. In addition, a study from Rodriguez-Lelis JM et al. discovered that the contact area between the femoral and tibial insert components had influenced von Mises stress and contact pressure [28].

The high-flexion knee prostheses proposed a thickened femoral posterior and a wider tibial insert contact surface to lower wear and contact pressure. Table 2.4 shows

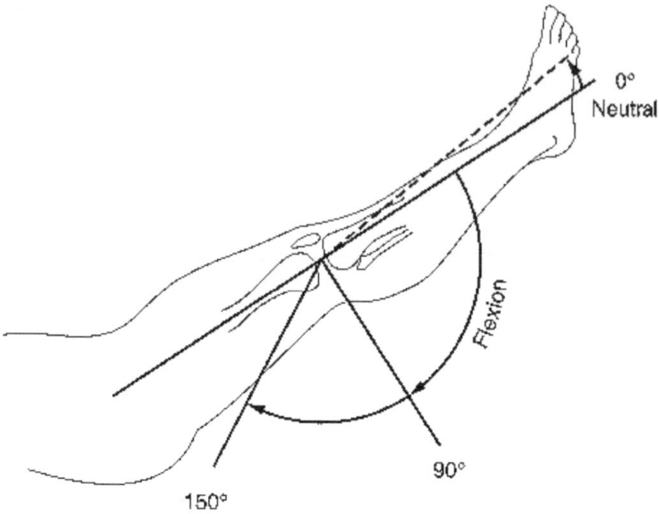

Fig. 2.9 Knee joint range of motion [26]

the features of high-flexion TKR for femoral and tibial insert parts. The NexGen LPS-Flex high-flexion total knee prosthesis from Zimmer Biomet, as shown in Fig. 2.10, is designed to support high-flexion activities. It proposed tibiofemoral contact with a high flexion of up to 155° minimising contact area loss in high flexion with a conforming femoral component and tibial insert design.

The design for high-flexion knee prostheses was proposed to lower contact stress and increase the contact area to achieve maximum flexion while maintaining stability throughout the range of motion. Table 2.5 shows the findings of high-flexion TKR for femoral and tibial insert parts over the years.

Table 2.4 Features of high-flexion knee prosthesis

Scientists, Year	Part	Findings	References
Kim et al. 2020	Femoral	The increase in thickness of the posterior femoral condyle by 2 mm may extend the flexion angle by 10° as the contact area increases in length between the posterior femoral and the tibial insert	[29]
Darmanto et al. 2022	Femoral	Adding 1 mm thickness to the posterior femoral condyle caused the flexion angle to increase by 3.2°	[30]
Triwardono et al. 2021	Tibial insert	Modified the tibial insert design by increasing the contact area to achieve flexion above 120°	[31]

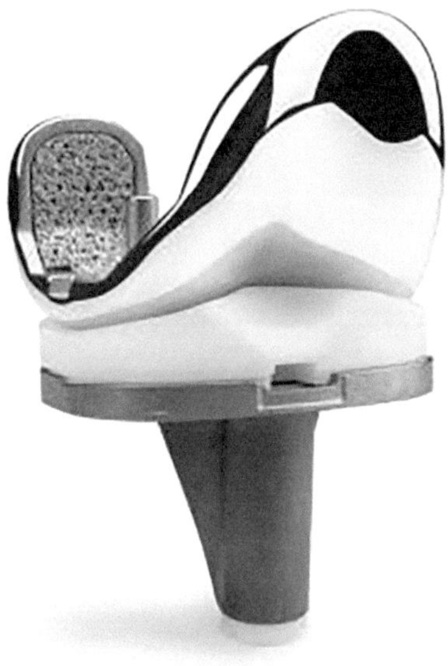

Fig. 2.10 NexGen LPS-Flex Knee [32]

Table 2.5 Studies on high-flexion knee prosthesis over the years

Scientists, Year	Findings	References
Long W et al. 2008	The main changes made to the component design are to lengthen the trochlear groove, adjust the cam-post design, recess the tibial insert, lengthen the radius of curvature via the posterior condyles, and increase the posterior condylar offset. These modifications enable more femoral translation, rollback, and clearance during deep flexion. Studies show that high flexion improves outcomes for both regular total knee and high-flexion designs	[33]
Jain S et al. 2013	The high-flexion knee prosthesis is made for people who need to bend their knees a lot, like those who often squat or kneel, which is common in Asians. Research shows it works well for some patients, offering similar results to standard knee replacements but allowing for more flexibility and better ability to squat and kneel	[34]
Sheldon M et al. 2018	This study compared the De-Puy posterior-stabilised rotating platform TKR prosthesis (PS-RP) with the more recent Hi-Flex PS-RPF device in terms of performance. The functional outcome and patient satisfaction of the two prostheses were specifically compared	[35]

Table 2.6 Research gap for thickness of tibial insert

Scientists, Year	Findings	References
Bartel et al. 1986	Maintaining a thickness of more than eight to ten millimetres should be the target for tibial insert components, according to the analysis of contact stress as a function of insert thickness for reducing the wear-related problem	[36]
El Deen et al. 2006	According to El-Deen's static tests, increasing the thickness of the polyethylene component resulted in a minor increase in contact imprint lengths [39]	[38]
Harish S et al. 2015	The simulation and experimental testing used three different thicknesses and sizes of tibial inserts. The large thickness and size reduced the deformation, von Mises stress, and maximum principal stress, enhancing the spacer's longevity and the implant	[40]
Y-G. Koh et al. 2019	The conforming enlarged design saw a smaller contact stress and a larger contact area. Increased conformance also led to a decrease in volumetric and wear rate wear. The enhanced conformance design, however, displayed constrained kinematics	[41]

2.5 The Different Thickness of Tibial Insert in TKR Components

Biomechanical studies have investigated how variations in tibial insert thickness affect knee joint mechanics. The thickness of the tibial insert component is one of the implant design criteria in the current TKR. A few publications address the connection between compressive pressure and the thickness of the tibial insert component in TKR [36–38]. Increasing the thickness of the tibial insert component would reduce contact pressure and increase the contact area, according to early research that used static finite element and elasticity simulations with modified knee implant geometry [36]. Table 2.6 shows several studies regarding the thickness of the tibial insert component throughout the years.

2.6 Biocompatibility Materials for TKR

A biocompatible material in medical implants can be defined as any material used to produce an implant that is safe and compatible to replace a functional body part [42]. Furthermore, the increased demand for implant surgery leads to the growth of biomaterials technology and medical implants. Hence, biomaterials are tested for safety and effectiveness in tissue and animal studies before being used in human fracture fixation [43].

The three components of knee prostheses require different materials and properties to function inside the body. Furthermore, the materials must not be highly stiff for the metallic femoral and tibial components, as the bone grows when it receives an acceptable amount of mechanical stimulation. The bone will experience lower tensile or compressive loads if the material is extremely stiff, inhibiting bone growth [44]. As a result, it reduces bone density, known as stress shielding. Moreover, the surface of femoral components is commonly designed with a highly polished surface to serve as the articulating surface along with the tibial insert, as shown in Fig. 2.11. A shiny surface helps reduce friction by lessening adhesion between surfaces and reducing the normal contact surface and shear stresses [44].

Common biomaterials for metallic femoral and tibial components are stainless steel, cobalt-chromium (Co-Cr) alloys, and titanium alloys [42]. However, titanium alloys are preferable to cobalt alloys and stainless steel. It is due to having lower modulus, better corrosion resistance properties, and better biocompatibility [46], as shown in Table 2.7. Since titanium releases very few metallic ions, many scientists consider it a bioinert material, making it more biocompatible than stainless steel or cobalt-chromium TKA implants [47].

Table 2.8 shows the properties of materials for stainless steel, cobalt-base alloys, and titanium alloys used for biomaterial implants.

The applications of polymers vary in tissue repair, replacement, and wound healing. Hence, polymers have many properties in the medical field, including lower density, high strength, increased durability, and sometimes lower cost [48]. Additionally, the biocompatible materials for the tibial insert component used polymers to replace the damaged cartilage. Polyethylene has been applied extensively as a biomaterial in medical implants, such as porous high-density polyethylene for facial

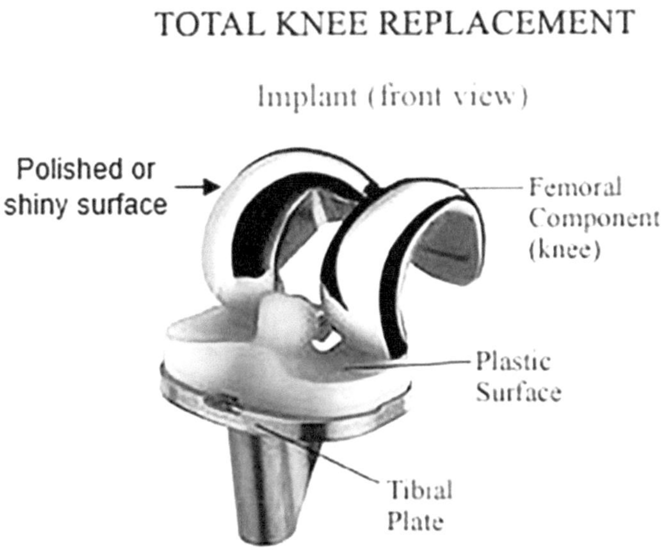

Fig. 2.11 Polished surface on the femoral components [45]

Table 2.7 Pros and cons of implant materials [46]

Materials	Pros	Cons
Stainless Steel	• Low Cost • High Availability • Processing	• Long-term behaviour • High modulus
Cobalt-base Alloys	Wear resistance Corrosion resistance Fatigue strength	High modulus Short term biocompatibility
Titanium Alloys	• Long-term biocompatibility • Corrosion resistance • Low modulus • Fatigue strength	• Lower wear resistance • Low shear strength

Table 2.8 Properties of implant materials [46]

Materials	Young Modulus, *E* (GPa)	Yield Strength (MPa)	Ultimate Tensile Strength (MPa)	Poisson Ratio, ν
Stainless Steel	200	170–750	465–950	0.3
Cobalt-base Alloys	200–230	275–1585	600–1795	0.3
Titanium Alloys	110	850–930	960–970	0.3

reconstruction [49]. Meanwhile, polyamides have been used widely as a biomaterial in medical applications such as sutures, catheters, dentures, and many more [50]. Besides, the development of polyamide 12 as the new biomaterial to replace polyethylene is introduced and utilised in medical implants. Table 2.9 shows the pros and cons of using polyethylene and polyamide 12 as biomaterials for tibial insert in TKR.

Table 2.9 Pros and cons of biomaterials in TKR

Materials	Pros	Cons	References
Polyethylene (PE)	• Wear resistance • Low friction • Biocompatibility • Fatigue strength • High impact strength	Wear debris released over a prolonged period	[51]
Polyamide 12 (PA 12)	• Wear resistance • Lower friction than PE results in reduced wear • Biocompatibility • High impact strength • Lightweight	Newer development, hence the long-term clinical data may be limited	[52]

Table 2.10 Material properties of polyethylene and polyamides

Materials	Density (kg/m^3)	Young's Modulus (MPa)	Yield Strength (MPa)	Poisson Ratio	References
Polyethylene	800–930	820–910	20–43	0.46	[53, 54]
Polyamide 12	1000–1020	1701	34.8–55	0.39	[50, 55]

Table 2.10 shows the material properties of polyethylene and polyamide 12 used in the tibial insert of TKR.

Nevertheless, polyamide 12 has the most potential for application in biomedical implants because of its outstanding mechanical properties and excellent biocompatibility [56]. In addition, polyamide 12 has an advantage over other biomedical polymers as it has been discovered to be an excellent material in fused filament fabrication (FFF) 3D printing [56]. Moreover, a study shows that polyamide 12 was chosen over other polyamides for its lower processing temperature, which is preferable in creating prototypes and parts [57].

2.7 Summary

The human knee joint anatomy and TKR have been briefly introduced. The designs of knee prostheses and high-flexion knee prostheses have been reviewed in this chapter. The biocompatibility material for TKR was addressed. The application of finite element analysis in TKR was discussed as well. FEA has been widely implemented on the TKR in previous studies regarding loading and boundary conditions. FEA outcomes can vary in terms of total deformation and stress distribution on the TKR. The preparation of jigs and compression testing for TKR has been extensively discussed, as well as the fabrication process for knee prostheses. Therefore, the next chapter should address the procedure of design selection, materials applied for knee prosthesis, the boundary conditions for FEA implemented on the TKR, and the preparation of jigs for compression testing.

References

1. T. Nagura, C.O. Dyrby, E.J. Alexander, T.P. Andriacchi, Mechanical loads on the knee joint during deep flexion. J. Orthopaedic Res. **20**(4), 881–886 (2002)
2. M.T. Hirschmann, W. Müller, Complex function of the knee joint: The current understanding of the knee. Knee Surgery, Sports Traumatol. Arthroscopy **23**(10), 2780–2788 (2015)
3. L. A. Pruitt, A.M. Chakravartula, *Mechanics of Biomaterials: Fundamental Principles for Implant Design*. Cambridge University Press (2011)
4. S.T. Lovald, K.L. Ong, E.C. Lau, J.K. Schmier, K.J. Bozic, S.M. Kurtz, Mortality, cost, and health outcomes of total knee arthroplasty in medicare patients. J. Arthroplasty **28**(3), 449–454 (2012)

5. G.M. Martin, I. Harris, Complications of total knee arthroplasty. Wolters Kluwer: UpToDate, Mar. 2022. [Online]. Available: https://www.uptodate.com/contents/complications-of-total-knee-arthroplasty#H1130979836 [Accessed 27 Feb 2023
6. P. Vivian Papas, F.D. Cushner, G.R. Scuderi, The history of total knee arthroplasty. Techn. Orthopaedics **33**(1), 2–6 (2018)
7. C.S. Ranawat, History of total knee replacement. J. Southern Orthopaedic Assoc. **11**(4), 218–226 (2002)
8. B.C. Carr, T. Goswami, Knee implants—review of models and biomechanics. Mater. Design **30**, 398–413 (2009)
9. I.S. Oh, M.K. Kim, D.S. You, S.B. Kang, K.H. Lee, Total knee arthroplasty without patellar resurfacing. Int. Orthopaedics **30**(5), 415–419 (2006)
10. J. Zhang, E. Li, Y. Zhang, Prostheses option in revision total knee arthroplasty, from the bench to the bedside: (1) basic science and principles. EFORT Open Rev. **7**(2), 174–187 (2022)
11. M.I. Ellen, D.R. Forbush, T.E. Groomes, Total knee arthroplasty, in Es*sentials of Physical Medicine and Rehabilitation*, 4th ed., W.R. Frontera, J. K. Silver, T.D. Rizzo, Philadelphia: Elsevier Inc., pp. 443–450 (2020)
12. E.K. Song, J.K. Seon, J.Y. Moon, J.H. Yim, The evolution of modern total knee prostheses, in *Arthroplasty—Update*, P. Kinov, UK: InTech Open (2013)
13. C. Dall'Oca et al., Evolution of TKA design. Acta Biomedica **88**, 17–31 (2017)
14. M. Karadsheh, TKA prosthesis design. Orthobullets (2021). [Online]. Available: https://www.orthobullets.com/recon/5019/tka-prosthesis-design. Accessed 1 Jun 2022
15. NexGen Complete Knee Solution, "Surgical Technique: NexGen ® Cruciate Retaining (CR) and Revision Instrumentation," Zimmer Biomet (2017). [Online]. Available: https://www.zimmerbiomet.eu/en/products-and-solutions/specialties/knee/nexgen-complete-knee-solution.html. Accessed 23 Oct 2020
16. R. D'Anchise, M. Andreata, C. Balbino, N. Manta, Posterior cruciate ligament-retaining and posterior-stabilized total knee arthroplasty: differences in surgical technique. Joints **1**(1), 5–9 (2013)
17. Joint Replacement Patient Advocacy and Online Community, "MicroPort orthopedics," Bone Smart (2013). [Online]. Available: https://bonesmart.org/sponsors/microport/?source=imagecaption. Accessed 23 Oct 2020
18. J.S. Broberg, S. Ndoja, S.J. MacDonald, B.A. Lanting, M.G. Teeter, Comparison of contact kinematics in posterior-stabilized and cruciate-retaining total knee arthroplasty at long-term follow-up. J. Arthroplasty **35**(1), 272–277 (2020)
19. C. Jiang, Z. Liu, Y. Wang, Y. Bian, B. Feng, X. Weng, Posterior cruciate ligament retention versus posterior stabilization for total knee arthroplasty: a meta-analysis. PLoS One **11**(1) (2016)
20. F.R. Kolisek et al., Posterior-stabilized versus posterior cruciate ligament-retaining total knee arthroplasty. Iowa Orthopaedic J. **29**, 23–27 (2009)
21. F.F. Buechel, F.F. Buechel, M.J. Pappas, J.D. Alessio, Twenty-year evaluation of meniscal bearing and rotating platform knee replacements. Clinical Orthopaedics Related Res. **388**, 41–50 (2001)
22. R.S. Laskin, The genesis total knee prosthesis: a 10-year follow up study. Clin. Orthopaedics Related Res. **388**, 95–102 (2001)
23. M.J. Cromie, R.A. Siston, N.J. Giori, S.L. Delp, Posterior cruciate ligament removal contributes to abnormal knee motion during posterior stabilized total knee arthroplasty. J. Orthopaedic Res. **26**(11), 1494–1499 (2008)
24. C. Browne, J.C. Hermida, A. Bergula, C.W. Colwell, D.D. D'Lima, Patellofemoral forces after total knee arthroplasty: effect of extensor moment arm. Knee **12**(2), 81–88 (2005)
25. D.L. Churchill, S.J. Incavo, C.C. Johnson, B.D. Beynnon, The influence of femoral rollback on patellofemoral contact loads in total knee arthroplasty. J. Arthroplasty **16**(7), 909–918 (2001)
26. N. Hamilton, W. Weimar, K. Luttgens, *Kinesiology: Scientific Basis of Human Motion*, 12th ed. McGraw Hill (2011)

27. H.T. Huang, Y.S. Jiing, G.J. Wang, The early results of high-flex total knee arthroplasty: a minimum of 2 years of follow-up. J. Arthroplasty **20**(5), 674–679 (2005)
28. J.M. Rodríguez-Lelis, J.A. Arellano-Cabrera, A. Abúndez-Pliego, J. Pérez Ortega, A numerical analysis of the contact stresses on surface of the femoral condyles of a knee prosthesis. Adv. Tissue Eng. Regenerative Med. **7**(1), 28–33 (2021)
29. Y.H. Kim, J.W. Park, Comparison of modular conventional and high-flexion posterior-stabilized total knee arthroplasties in the same patients at a mean follow-up of 15 years. J. Arthroplasty **35**(5), 1262–1267 (2020)
30. D. Darmanto, R. Novriansyah, P.W. Anggoro, R. Ismail, J. Jamari, A.P. Bayuseno, A review on flexion angle in high-flexion total knee arthroplasty for Indonesian's need. Frontiers Mech. Eng. **8** (2022)
31. J. Triwardono et al., Evaluation of the contact area in total knee arthroplasty designed for deep knee flexion. Int. J. Technol. **12**(6), 1312–1322 (2021)
32. NexGen® Complete Knee Solution | Proven, "Supports High-Flexion Activities," Zimmer Biomet. [Online]. Available: https://www.zimmerbiomet.lat/en/medical-professionals/knee/product/nexgen-complete-knee-solution.html. Accessed 31 Jul 2021
33. W.J. Long, G.R. Scuderi, High-flexion total knee arthroplasty. J. Arthroplasty **23**(7), 6–10 (2008)
34. S. Jain, A.C. Pathak, K. Kanniyan, S. Kulkarni, S. Tawar, P. Mane, High-flexion posterior-stabilized total knee prosthesis: Is it worth the hype?. Knee Surgery Related Res. **25**(3), 100–105 (2013)
35. M. Sheldon, S. Ryan Du, Y. Piers, Performance of hi-flex femoral component in total knee arthroplasty—a randomized control study'. J. Disorders Treatment **4**(2) (2018)
36. D.L. Bartel, V.L. Bicknell, T.M. Wright, The effect of conformity, thickness, and material on stresses in ultra-high molecular weight components for total joint replacement. J. Bone Joint Surgery Amer. **68**, 1041–1051 (1986)
37. Y. Bei, B.J. Fregly, W.G. Sawyer, S.A. Banks, N.H. Kim, The relationship between contact pressure, insert thickness, and mild wear in total knee replacements. Tech Sci. Press CMES **6**(2), 145–152 (2004)
38. M. EI-Deen, M. Garcia-Finana, Z.-M. Jin, Effect of ultra-high molecular weight polyethylene thickness on contact mechanics in total knee replacement. Proc. Inst. Mech. Eng. **220**(7), 733–742 (2006)
39. G.S. Matharu, D.J. Culliford, A.W. Blom, A. Judge, Projections for primary hip and knee replacement surgery up to the year 2060: an analysis based on data from The National Joint Registry for England, Wales, Northern Ireland and the Isle of Man. Annals Royal College Surgeons England **104**(6), 443–448 (2022)
40. S. Harish, V.R. Devadath, Additive manufacturing and analysis of tibial insert in total knee replacement implant. Int. Res. J. Eng. Technol. **2**(4), 633–638 (2015)
41. Y.-G. Koh, K.-M. Park, H.-Y. Lee, K.-T. Kang, Influence of tibiofemoral congruency design on the wear of patient-specific unicompartmental knee arthroplasty using finite element analysis. Bone Joint Res. **8**(3), 156–164 (2019)
42. U.S. Gupta, A. Patel, V. Gehlot, M.L. Jain, A review of biocompatible materials for medical implants using additive manufacturing. Int. J. Adv. Res. Sci. Eng. **7**(5), 252–263 (2018)
43. M.R. Cohn, A. Unnanuntana, T.J. Pannu, S.J. Warner, J.M. Lane, Materials in fracture fixation. Comprehensive Biomaterials II **7**, 278–297 (2016)
44. M. Silva, K. Snow, Biotribology considerations of total knee replacement. Dept. Mecha. Indus. Eng., Northeastern Univ., Boston, MA, USA (2009)
45. Douglas, Which Type of Implant is Right For You?. kneereplacementcosts.com (2021) [Online]. Available: https://www.kneereplacementcosts.com/implant-types.html. Accessed 12 Apr 2022
46. M. Long, H.J. Rack, Titanium alloys in total joint replacement-a materials science perspective. Biomaterials **19**(18), 1621–1639 (1998)
47. N. Eliaz, Corrosion of metallic biomaterials: a review. Materials **12**(3) (2019)

48. M. Soroush, M.C. Grady, Polymers, polymerization reactions, and computational quantum chemistry, in *Computational Quantum Chemistry*, M. Soroush, Elsevier Inc., pp. 1–16 (2019)
49. N.C. Paxton, M.C. Allenby, P.M. Lewis, M.A. Woodruff, Biomedical applications of polyethylene. Europ. Polymer J. **118**, 412–428 (2019)
50. M. Shakiba et al., Nylon—A material introduction and overview for biomedical applications. Polymers Adv. Technol. **32**(9), 3368 (2021)
51. G. Chakrabarty, M. Vashishtha, D. Leeder, Polyethylene in knee arthroplasty: a review. J. Clin. Orthopaedics Trauma **6**(2), 108–112 (2015)
52. G.F. Delgado, A.C. Pinho, A.P. Piedade, 3D printing for cartilage replacement: a preliminary study to explore new polymers. Polymers (Basel) **14**(5) (2022)
53. X.H. Wang, Z. Bi, X. Dong, S.Y. Hu, L. jun Xu, C.K. Cheng, Increasing the height of the anterior lip on a tibial insert in a posterior stabilized knee prosthesis has little effect on the wear rate. Med. Eng. Phys. **91**, 48–53 (2021)
54. L. Zach, L. Kunčická, P. Růžička, R. Kocich, Design, analysis and verification of a knee joint oncological prosthesis finite element model. Comput. Biol. Med. **54**, 53–60 (2014)
55. D.A. Porter, M.A. Di Prima, Y. Badhe, A.R. Parikh, Nylon lattice design parameter effects on additively manufactured structural performance. J. Mech. Behav. Biomed. Mater. **125** (2022)
56. T.N.A.T. Rahim, A.M. Abdullah, H.M. Akil, D. Mohamad, Comparison of mechanical properties for polyamide 12 composite-based biomaterials fabricated by fused filament fabrication and injection molding, in *AIP Conference Prococeedings*, American Institute of Physics Inc. (2016)
57. T.N.A.T. Rahim, A.M. Abdullah, H.M. Akil, D. Mohamad, Z.A. Rajion, The improvement of mechanical and thermal properties of polyamide 12 3D printed parts by fused deposition modelling. Express Polymer Lett. **11**(12), 963–982 (2017)

Chapter 3
Finite Element Analysis (FEA) on Knee Prosthesis

3.1 Finite Element Analysis (FEA) on TKR

The broad implementation of finite element analysis has been one of the most significant improvements since the 1950s. Currently, the design of products and systems uses FEA as the standard analysis tool. FEA can be described as using numerical computation, mathematical calculations, models, and simulations to solve, analyse, and estimate the behaviour of an object, part, assembly, or structure under several physical conditions. Engineers frequently used the application of FEA to simulate physical forces to find design flaws in their product prototypes. Furthermore, FEA simplifies the overall design for engineers to focus on by enabling specific design issues and optimising individual parts.

Most studies on TKR use FEA to examine further fatigue, wear, deformations, stresses, and pressure in knee prostheses and analyse the reasons for implant failures. Therefore, finite elements used by previous scientists on the TKR models are listed in Table 3.1.

3.2 Loading and Boundary Conditions of TKR

One of the crucial factors in implementing finite element analysis is applying loading and boundary conditions to the 3D model. This element must be assigned to accurately represent or mimic the knee joint's motion in daily activities. However, setting loading and boundary conditions can be challenging because of the complicated interaction between the bones, muscles, ligaments, and tendons in diverse human movements [5]. Besides that, this complicated interaction enables the knee to tolerate significant stresses throughout various daily activities [6]. Therefore, applying loading and knee kinematics is based on assumptions of forces and pressures acting in the human body assigned as the boundary conditions to the finite

I. M. Arrif et al., *High-Flexion Knee Prosthesis*,
SpringerBriefs in Applied Sciences and Technology,
https://doi.org/10.1007/978-981-96-9878-3_3

Table 3.1 List of finite elements used by previous scientists on the TKR

Scientists, Year	Main objective	Outcomes of FEA	References
Sun et al., 2015	To study the implanted in vivo contact mechanics of the customised knee prosthesis	The study analysed total deformation at the femur and contact pressure at the tibial plateau using FEA	[1]
Li et al., 2017	To develop customised total knee replacement implants	Predicted wear on polyethylene insert of customised TKR using finite element model	[2]
Mou et al., 2018	To ensure the precise implantation of a femoral component by optimising and customising the component's placement configurations	The study analysed the pressure on the polyethylene liner using FEA	[3]
Shuib et al., 2021	To create a modified knee implant design capable of flexing up to 165°	Compared and evaluated modified designs in terms of total deformation, von Mises stress, shear stress, and contact pressure on knee implants using FEA	[4]

element model. For instance, the force applied to the femur bone as loading condition and fixed contact to the tibia as boundary conditions were applied to the FE knee model, as shown in Fig. 3.1.

In Malaysia, the average body weight considered was 652.365 N taken based on 94 patients with osteoarthritis who underwent total knee replacement surgery [8]. The average weight was equivalent to 66.5 kg applied vertically downward towards the Z-axis, as shown in Fig. 3.1. Moreover, the angle of flexion is introduced to represent when the knee joint is straightened and bent. Additionally, it is measured in degrees, with 0° indicating complete extension and 180° expressing full bending. However, the applied forces depend on the percentage of body weight and angle of flexion to recreate knee load during daily activities. These applied forces were calculated from the percentage of body weight times the average body weight. The percentage of body weight and angle of flexion implemented on the FE models of TKR to represent daily activities are listed in Table 3.2 [9].

Fixed and contact constraints are introduced during FEA as boundary conditions implemented on 3D knee prostheses. Based on the literature [10], fixed support is applied on the tibial component, as shown in Fig. 3.2. Meanwhile, the tibial insert is bonded to the tibial component, and the same bonded contact is applied between the femoral component and the tibial insert. The bonded connection modelled by the FE is a constraint between the bodies to prevent mutual rotations and movements [11].

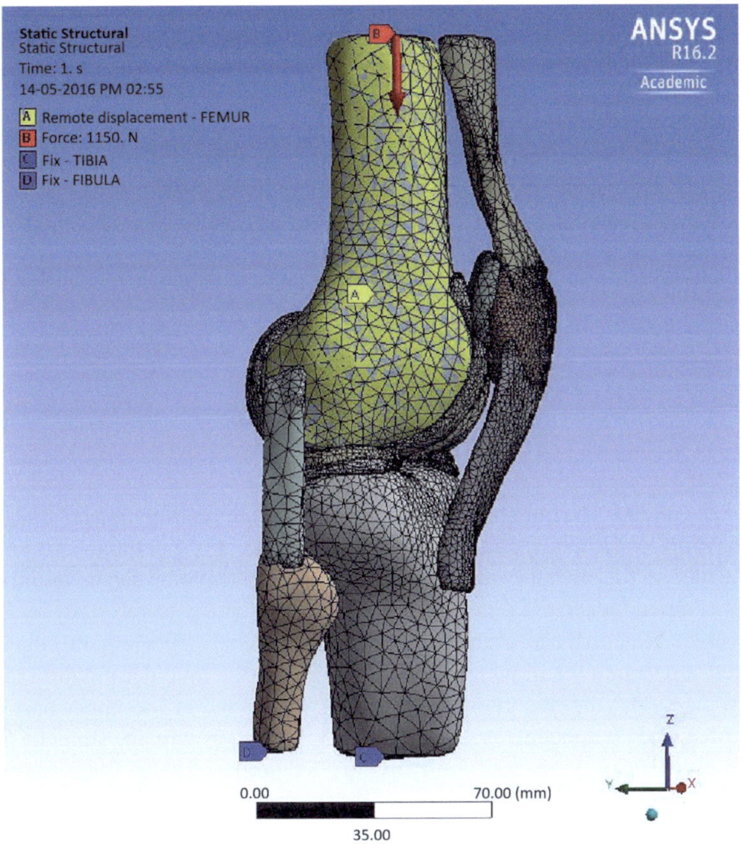

Fig. 3.1 Loading and boundary conditions applied on the knee joint [7]

Table 3.2 Percentage of body weight during daily activities on TKR [9]

Activities	Angle Flexion (°)	Percentage of body weight (%)
Standing	0	0.5
Kneeling	90	1.2
Squatting	135	5.3
Sitting on Feet	165	6.0

3.3 Total Deformation on TKR

Total deformation can be defined as the displacement and deformation that the TKR experiences due to the loading and boundary conditions during the simulation. It displays all the geometric changes caused by the applied forces and constraints in

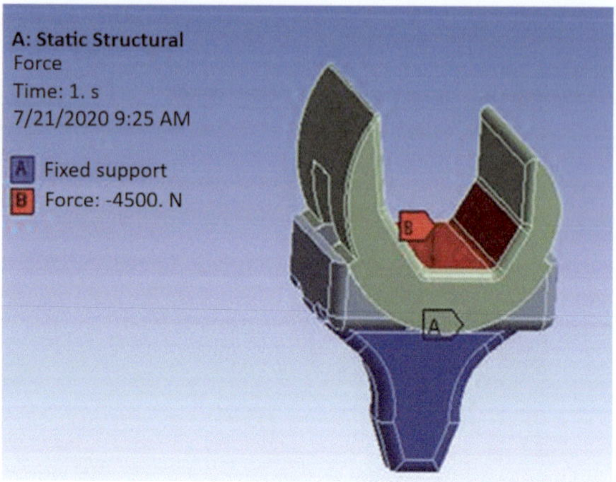

Fig. 3.2 Fixed support applied on the TKR [10]

the FE knee model. The directional deformation measures in directions like X, Y, and Z. Meanwhile, total deformation includes all three directions in the measurement.

Simulated as the assembly of TKR, usually total deformation can be found on the femoral component and tibial insert during FEA [12]. The research study by Kumar implemented FEA on the assembly TKR with load applied on the femoral component and fixed support on the tibial component to observe the total deformation of TKR [10]. In the study, the maximum deformation was found at the upper part of the anterior femoral component, as shown in Fig. 3.3. Furthermore, the study by Seman et al. observed the different conformity of the tibial insert when the force was applied to the different angle flexions of TKR [13]. The trend of total deformation can be described as inclining with increasing angle flexion.

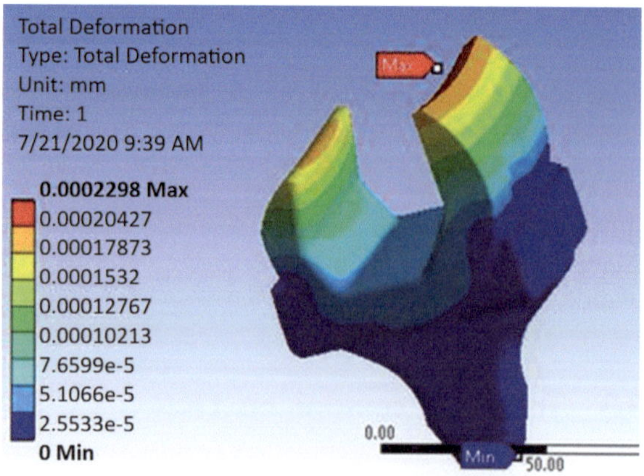

Fig. 3.3 Maximum total deformation on the TKR [10]

3.4 Stress Distribution on TKR

Stress distribution can be described as the internal stresses within a knee prosthesis and the bones and soft tissues in response to external loads and pressures. As stress is distributed in the static state, the TKR outcome is focused on von Mises stress and contact pressure. The von Mises stress determines if the assigned material to the TKR components will yield or fracture under applied load. Meanwhile, the contact pressure between two bodies develops pressure under the load applied to a contact area.

A few research studies have implemented FEA on the assembly TKR to analyse the stress distribution. For instance, the study by Kumar Madeti et al. used loading and fixed constraint to the femoral and tibial components, respectively, and implemented FEA on the assembly TKR to analyse the von Mises stress [12], and the most distributed equivalent stress was observed at the femoral condyles, as shown in Fig. 3.4.

Furthermore, the research by Ingrassia et al. utilised FEA on the assembly TKR to measure the contact stress or pressure on the tibial insert [11]. The study observed that the contact pressure was found at the medial, lateral, and posterior sections and the post of the tibial insert, as shown in Figs. 3.5 and 3.6. Meanwhile, Loi et al. used different boundary conditions but still measured the contact stresses on the tibial insert [14].

Moreover, the study by Seman et al. presented the different conformity of the tibial insert when the force was applied to the assembly of TKR [13]. The trend of contact pressure can be described as increasing against angle flexion, as shown in

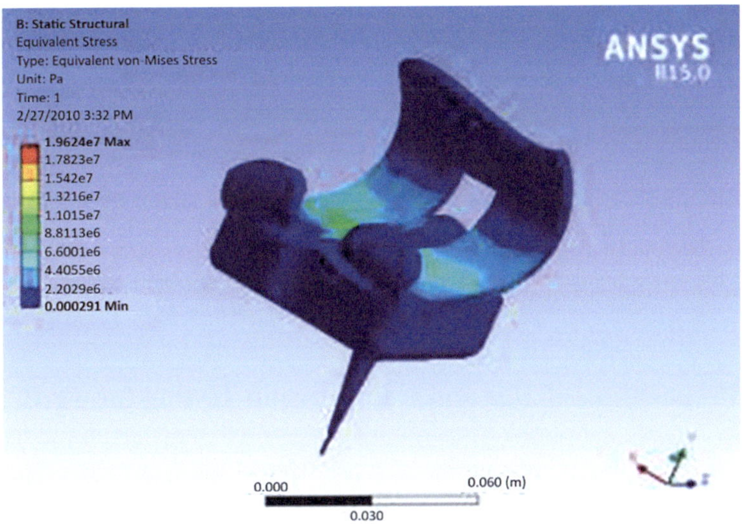

Fig. 3.4 Equivalent (von Mises) Stress on the TKR [12]

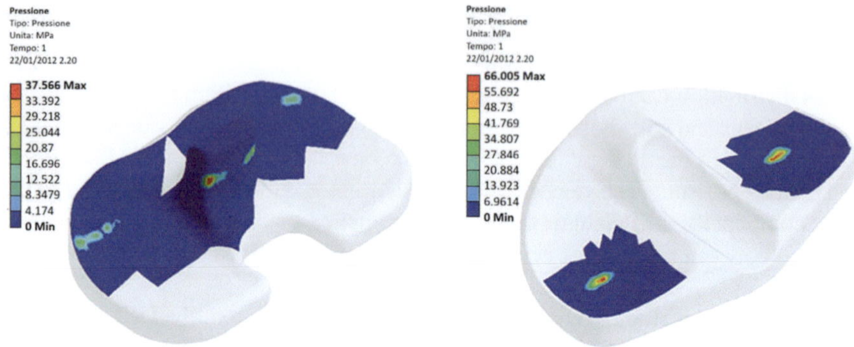

Fig. 3.5 Contact pressure on the TKR [11]

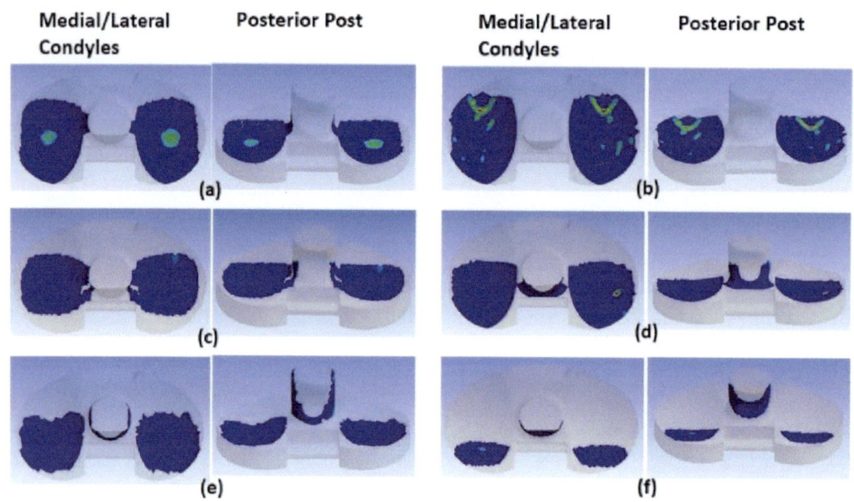

Fig. 3.6 Contact pressure on the TKR [15]

Fig. 3.7. The study also observed that the trend of von Mises stress was increased against the angle flexion with different forces applied to different angles.

3.5 Preparation of Jigs and Compression Testing for TKR

Compression testing on knee prostheses is part of mechanical testing used to determine the stiffness and elasticity of material when subjected to compressive forces. Mostly, standard testing is carried out involving the assembly of the TKR components. The organisations that manage testing methodologies are commonly known

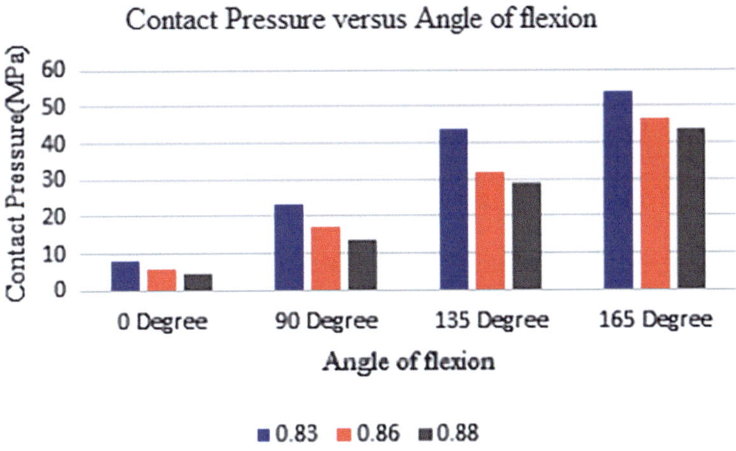

Fig. 3.7 Trend of contact pressure on the TKR [13]

as the American Society of Testing Materials (ASTM) and the International Organization for Standardization (ISO). Specifically, the compression testing referred to the standard of ASTM F2777 for evaluating tibial insert endurance and deformation under high flexion.

The compression test can be described as a specimen uniformly loaded with an increasing force in the longitudinal direction, starting with a low initial force [16]. The compressive Young's modulus, yield point, and ultimate compressive stress can all be calculated, as well as the modulus of elasticity, also known as Young's modulus. Young's modulus is the feature that defines the material's elasticity and is also known as the material's stiffness.

Furthermore, the advantages of conducting compression testing are discovering the forces, resulting deformation, and the contact pressure that ultimately causes component failure. The compression tests performed by previous scientists on the TKR are listed in Table 3.3.

The universal testing machine (UTM) is the most common machine used to conduct compression testing. The tensile stress and compressive strength of materials are tested using a UTM. The machine is adaptable and can be programmed to provide a wide range of loading profiles under different boundary circumstances. Figures 3.8 and 3.9 display the setup of experimental compression testing, including TKR components and fixtures.

The research by Suyitno et al. conducted compression testing on the assembly of TKR components to compare the displacement results with FEA, and the comparison results for both test and FEA show from 0.89% to 1.45% differences [18]. Moreover, the study by Sakai R. et al. presented a compression test conducted on the post-cam of knee implants, with the von Mises stress results compared between experimental testing and FEA. To conclude, the maximum von Mises stress results between experimental and FEA show a 15% difference [19].

Table 3.3 List of compression tests used by previous scientists on the TKR

Scientists, Year	Main objective	Outcomes of compression test	References
Ketata et al., 2010	To examine how the tibiofemoral contact response was affected by the total knee replacement design	Analysed load against displacement curves and compared the contact stiffness between two designs of tibial inserts	[17]
Suyitno et al., 2018	To create a precise method for determining the static compressive strength of an ultrahigh molecular weight knee joint component	Observed the displacement of the tibial insert during the compression test and compared the result with finite element simulation	[18]
Sakai R et al., 2019	To develop a feature for the tibial post present in the polyethylene inserts of posterior stabilise knee prostheses	Detected the surface pressure on the tibial insert post during the compression test and conducted finite element analysis at the tibial post in the posterior stabilised knee prosthesis	[19]

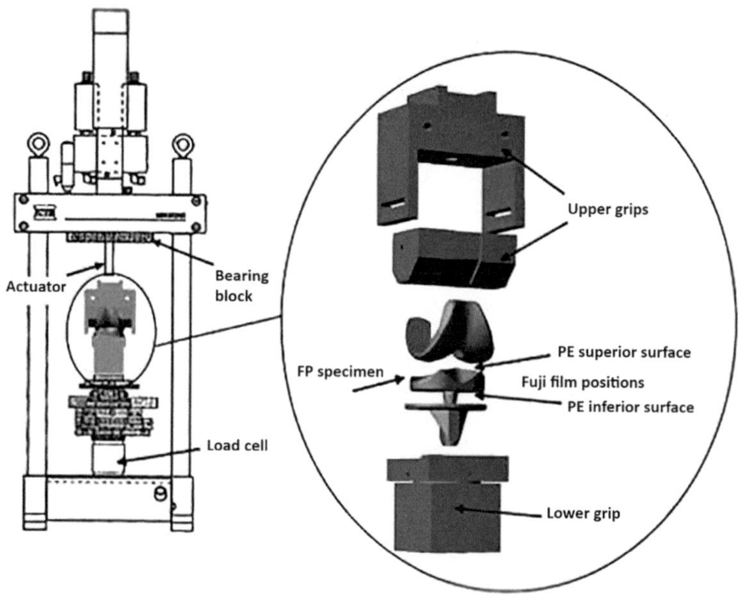

Fig. 3.8 Schematic diagram of compression testing setup [18]

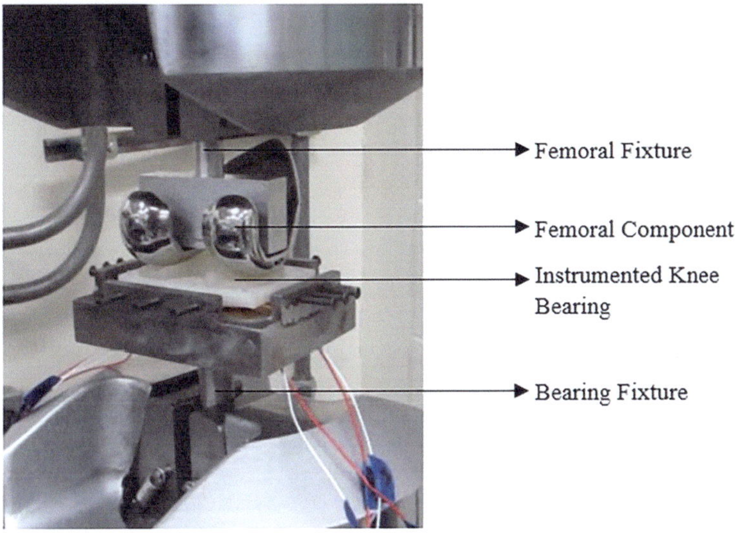

Fig. 3.9 Experimental compression testing setup [20]

3.6 Fabrication

Suitable fixtures or jigs to hold the TKR components are also needed during compression testing. Designs of jigs are developed by referring to the testing machine utilised when conducting experiments or testing. The UTM generally secured the femoral and tibial jigs and then applied load to the femoral component to produce loading. Usually, the material of the jigs depends on the material of the TKR components that enable the load to be transferred to the jigs through the femoral component of the TKR [21]. Lastly, the jigs are developed and completed with fabrication using computed numerical control (CNC) machining, classified as a subtractive manufacturing process.

Contrary to the fabrication of knee prostheses, which often uses additive manufacturing in nonengineering fields such as medicine, education, and prototype development before mass production, traditional manufacturing methods are still widely applied in large-scale production. The benefits of using AM in the medical field, especially in the orthopaedics, as described in Table 3.4.

Moreover, the AM process implemented on the TKR components is based on the material usage to be printed. Femoral and tibial components use metallic alloy, which is suitably printed using selective laser sintering (SLS). Meanwhile, the tibial insert used polymer material suitable for printing by fused deposition modelling (FDM). Table 3.5 shows the comparison of using SLS and FDM [26].

Table 3.4 Benefits of using additive manufacturing in orthopaedics

Applications	Description
Anatomic model for surgical planning and guide	Bone models can be produced from patient data by using AM technologies, which can help surgeons have a better plan before performing the surgery procedure [22]
Cost-effectiveness	The customised implants and tools using AM give a greater production rate and cheaper cost than the conventional system [23]
Patient-specific instruments	Specific instruments, such as drills and saws, can be fabricated by AM technologies, which ultimately reduces operation time [23]
3D-printed custom implants	The conventional manufacturing method produces standard-sized knee implants that do not accurately fit most patients. Thus, applying AM to customise knee implants that perfectly fit the knee joint can prevent knee implants from not fitting well into a patient's knee [24]
Patient care improvement	The custom implant produced by AM is more precise, reducing the operation risk with minimal invasion and lessening the patient's stress [25]

Table 3.5 Comparison of using SLS and FDM [26]

SLS	FDM
Used a high-intensity laser as an energy source to melt and fuse selective regions of powder layer by layer to create completely dense and complex metal components	Deposited the thermoplastic filament and extruded layer over layer through the heated nozzle to create a 3D object
Suitable for printing metals, including stainless steel, titanium alloy, nickel alloy, etc	Suitable for printing polymers such as ABS, PLA, PA, and more
Excels at printing complex and intricate shapes	Good at printing complex shapes but struggle with overhangs
More expensive and require specialised facilities	More affordable and accessible to a broader audience

3.7 Comparison Between Conventional TKR and Additive Manufacturing TKR

Additive manufacturing is a computer-controlled process that deposits materials layer by layer to create 3D objects. It uses computer-aided design (CAD) data or 3D object scanners to 3D printer machines to deposit material layer by layer in accuracy to form geometry shapes [27]. It contradicts traditional subtractive manufacturing, which removes material through machining, milling, or other means to create an object [28]. Hence, one advantage of additive manufacturing is that it reduces material waste, which reduces the cost of production and lessens the time taken to create a product. Therefore, it does not require a complex assembly. Besides, additive manufacturing can create complex geometries that are impossible using subtractive manufacturing methods. As a result, it allows the creation of customised parts in small batches and mass customisation in the future [29].

Various aspects can affect the cost comparison between knee implants made via additive manufacturing or 3D printing and conventional knee implants. At first glance, conventional TKR could seem more affordable because of its extensive availability and well-established production procedures. However, it may be less expensive in the long run because AM can create customised implants more quickly and with less material waste [30]. Table 3.6 shows the cost comparison between conventional TKR and additive manufacturing TKR [30, 32].

Table 3.6 Cost comparison between conventional TKR and additive manufacturing TKR

Aspects	Conventional TKR	Additive manufacturing TKR
Material cost	May vary depending on the type of metal or polymer used	Often utilises specialised materials such as titanium alloys or medical-grade polymers, which may have higher upfront costs but could be more efficient in terms of material utilisation
Manufacturing cost	Manufacturing conventional implants involves machining and other subtractive processes, which require labour, machinery, and tooling costs	It involves using 3D printers with associated costs such as equipment purchase, maintenance, and operator training. However, AM can be more cost-effective for complex geometries and customisation
Customisation cost	Conventional implant customisation often involves additional manual labour and may incur higher costs	AM enables mass customisation without significant additional costs, potentially making it more cost-effective for tailored implants
Waste and efficiency	Traditional manufacturing methods can generate more material waste due to subtractive processes	AM is generally more material-efficient, using only the necessary material for each implant
Lead times	Manufacturing lead times for conventional implants can be longer due to the complexity of machining and production processes	AM can offer shorter lead times, especially for customised implants, potentially reducing inventory costs and improving patient outcomes
Complexity	Complex geometries are challenging or impossible to achieve with conventional manufacturing methods	AM enables the creation of complex geometries and structures. While this may increase the upfront design and production cost, it can improve implant performance and longevity
Supply Chain	Conventional knee implants may have a more established supply chain, resulting in lower costs due to economies of scale	Especially for customised implants, AM may require more specialised materials and processes, which could initially increase costs

Ultimately, several changing elements, such as the patient's specific requirements, the production volume, the degree of customisation required, and the effectiveness of the manufacturing process, will determine whether conventional or AM implants are more cost-effective. Although there are situations where conventional implants are less expensive initially, AM has the potential to reduce costs over time while also improving patient outcomes and product performance [31].

3.8 Summary

The human knee joint anatomy and TKR have been briefly introduced. The design of knee prostheses and high-flexion knee prostheses has been reviewed in this chapter. The biocompatibility material for TKR was addressed. The application of finite element analysis in TKR was discussed as well. FEA has been widely implemented on the TKR in previous studies regarding loading and boundary conditions. FEA outcomes can vary in terms of total deformation and stress distribution on the TKR. The preparation of jigs and compression testing for TKR has been extensively discussed, as well as the fabrication process for knee prostheses. Therefore, the next chapter should address the procedure of design selection, materials applied for knee prosthesis, the boundary conditions for FEA implemented on the TKR, and the preparation of jigs for compression testing.

References

1. C. Sun et al., Finite element analysis of a retrieved custom-made knee prosthesis. J. Mech. Med. Biol. **15**(3) (2015)
2. X. Li, C. Wang, Y. Guo, W. Chen, An approach to developing customized total knee replacement implants. J. Healthcare Eng. **2017**, 1–8 (2017)
3. Z. Mou et al., Optimization of parameters for femoral component implantation during TKA using finite element analysis and orthogonal array testing. J. Orthopaedic Surgery Res. **13**(1) (2018)
4. S. Shuib, M.A. Azemi, I. Binti, M. Arrif, N.S. Hamizan, Design for additive manufacturing and finite element analysis for high flexion total knee replacement (TKR). J. Mech. Eng. **18**(2), 97–110 (2021)
5. B.K. Madeti, S.R. Chalamalasetti, S.K.S. Siva Rao Bolla Pragada, Biomechanics of knee joint—a review. Frontiers Mech. Eng. **10**(2), 176–186 (2015)
6. A. Chhabra, C. Curtis Elliott, M.D. Miller, Normal anatomy and biomechanics of the knee. Sports Med. Arthroscopy Rev. **9**(3), 166–177 (2001)
7. A.K. Devaraj et al., Finite element analysis of a human knee joint. *NAFEMS India Regional Conference*, Bangalore, pp. 1–8 (2016)
8. M.K. Che Hasan, Physical health and cardiometabolic risk markers among the elder people **5**(1), 34–39 (2017)
9. T. Nagura, C.O. Dyrby, E.J. Alexander, T.P. Andriacchi, Mechanical loads on the knee joint during deep flexion. J. Orthopaedic Res. **20**(4), 881–886 (2002)

10. M. Venkata Pavan Kumar, I. Bhanulatha, Design and structural analysys of knee joint using different materials with finite element method. Int. J. Adv. Scientific Res. Eng. Trends **5**(7), 32–38 (2020)
11. T. Ingrassia, L. Nalbone, V. Nigrelli, D. Tumino, V. Ricotta, Finite element analysis of two total knee joint prostheses. Int. J. Interactive Design Manuf. **7**(2), 91–101 (2013)
12. B. Kumar Madeti, C. Srinivasa Rao, Finite element analysis for knee implants with suitable material combinations. Int. J. Exp. Comput. Biomech. **4**, 113–124 (2018)
13. M.H. Seman, S. Shuib, M. Afzan, A.A. Shokri, Design and analysis of high flexion femoral component for total knee arthroplasty (TKA). Int. J. Eng. Technol. **7**, 381–385 (2018)
14. I. Loi, D. Stanev, K. Moustakas, Total knee replacement: subject-specific modeling, finite element analysis, and evaluation of dynamic activities. Frontiers Bioeng. Biotechnol. **9** (2021)
15. J. Triwardono et al., Evaluation of the contact area in total knee arthroplasty designed for deep knee flexion. Int. J. Technol. **12**(6), 1312–1322 (2021)
16. E. Friis, *Mechanical Testing of Orthopaedic Implants*. Woodhead Publishing Series (2017)
17. H. Ketata, A. Krichen, M. Dammak, A. Maalej, Design effect on the mechanical response of Total Knee Replacement under high-compression loading. Int. J. Biomed. Eng. Technol. **4**(1), 65–77 (2010)
18. Suyitno, E. Purba, L. Pudjilaksono, Experiment and FEM simulation of static test of knee joint prostheses, in *AIP Conference Proceedings*, American Institute of Physics Inc. (2018)
19. R. Sakai, U. Ken, K. Yoshida, K. Mabuchi, M. Ujihira, Compression test and finite element analysis for failure criterion of the tibial post of a posterior-stabilized knee prosthesis. J. Biomed. Sci. Eng. **12**(5), 285–292 (2019)
20. M. Safaei, R.M. Meneghini, S.R. Anton, Force detection, center of pressure tracking, and energy harvesting from a piezoelectric knee implant. Smart Mater. Struct. **27**(11) (2018)
21. C. Clary, L. Maletsky, *Mechanical Testing of Orthopaedic Implants*. Woodhead Publishing Series, pp. 207–229 (2017)
22. E. Walkenhorst, M. Talaat, A. Joynes, Manufacturing total knee replacement using 3D printing description of reconstructive (2019)
23. M. Javaid, A. Haleem, Additive manufacturing applications in orthopaedics: a review. J. Clin. Orthopaedics Trauma **9**(3), 202–206 (2018)
24. T.M. Wong et al., The use of three-dimensional printing technology in orthopaedic surgery: a review. J. Clin. Orthopaedics Trauma **25**(1) (2017)
25. B. Boonen, M.G.M. Schotanus, N.P. Kort, Preliminary experience with the patient-specific templating total knee arthroplasty. Acta Orthopaedica **83**(4), 383–393 (2012)
26. T. John Erinle, K.E. Ojaomo, O.S. Adeoye, Fused deposition modelling (FDM) and selective laser sintering (SLS) as indisputable machines in additive manufacturing: a review. *International Conference on 3D Printing & Additive Manufacturing*, pp. 1–9 (2023)
27. TWI. "What is Additive Manufacturing? Definition and Processes," TWI Ltd (2024). [Online]. Available: Available: https://www.twi-global.com/technical-knowledge/faqs/what-is-additive-manufacturing. Accessed 23 Oct 2020
28. GE Additive. "What is Additive Manufacturing?," GE Additive (2023). [Online]. Available: Available: https://www.ge.com/additive/additive-manufacturing. Accessed 23 Oct 2020
29. Mohd. Javaid, A. Haleem, Additive manufacturing applications in medical cases: a literature based review. Alexandria J. Med. **54**(4), 411–422 (2018)
30. A. Ariz, I. Tasneem, D. Bharti, A. Vaish, A. Haleem, M. Javaid, Is additive manufacturing of patient-specific implant is beneficial for orthopedics. Apollo Med (2021)
31. M. Le Stum et al., Three-dimensional printed knee implants: insights into surgeons' points of view. J. Personalized Med. **13**(5) (2023)
32. S. Harish, V.R. Devadath, Additive manufacturing and analysis of tibial insert in total knee replacement implant. Int. Res. J. Eng. Technol. **2**(4), 633–638 (2015)

Chapter 4
Development of High-Flexion TKR

4.1 Introduction

The procedure ensured that all research objectives were achieved and followed in a coherent sequence. The research framework begins with developing designs of the TKR components by identifying the existing designs of knee prostheses. Then, CAD software was used to proceed with the drawing process and design modifications of components involved in knee prostheses. The finite element analysis was carried out on the modified TKR components using FEA software. Assembly of TKR components for each modified design was simulated under various angle flexion, including 0°, 90°, 135°, and 165°. The finalised modified design of TKR was fabricated using additive manufacturing by outsourcing to 3D Gens Sdn Bhd. Lastly, compression tests were conducted on the printed TKR components by designing and fabricating suitable jigs to fit in the testing machine. Afterwards, the testing results on the TKR components were compared with the FEA results. As shown in Fig. 4.1, the flowchart best describes all these methods to achieve the research objectives.

4.2 Design Selection

As a simple design tool, the Pugh chart was applied in the design selection process for comparing design ideas against desired design criteria. The Pugh chart assists in decision-making as it converts qualitative decisions to quantitative information by listing various options. From a list of available options, the most crucial criteria are identified first to decide among the best options before the best and most practical option is chosen based on the requirements.

Table 4.1 shows the Pugh chart and the criteria for deciding the suitable type of design for total knee replacement. In this process, studying existing and standard designs for TKR was necessary before determining which type is suitable to place in

I. M. Arrif et al., *High-Flexion Knee Prosthesis*,
SpringerBriefs in Applied Sciences and Technology,
https://doi.org/10.1007/978-981-96-9878-3_4

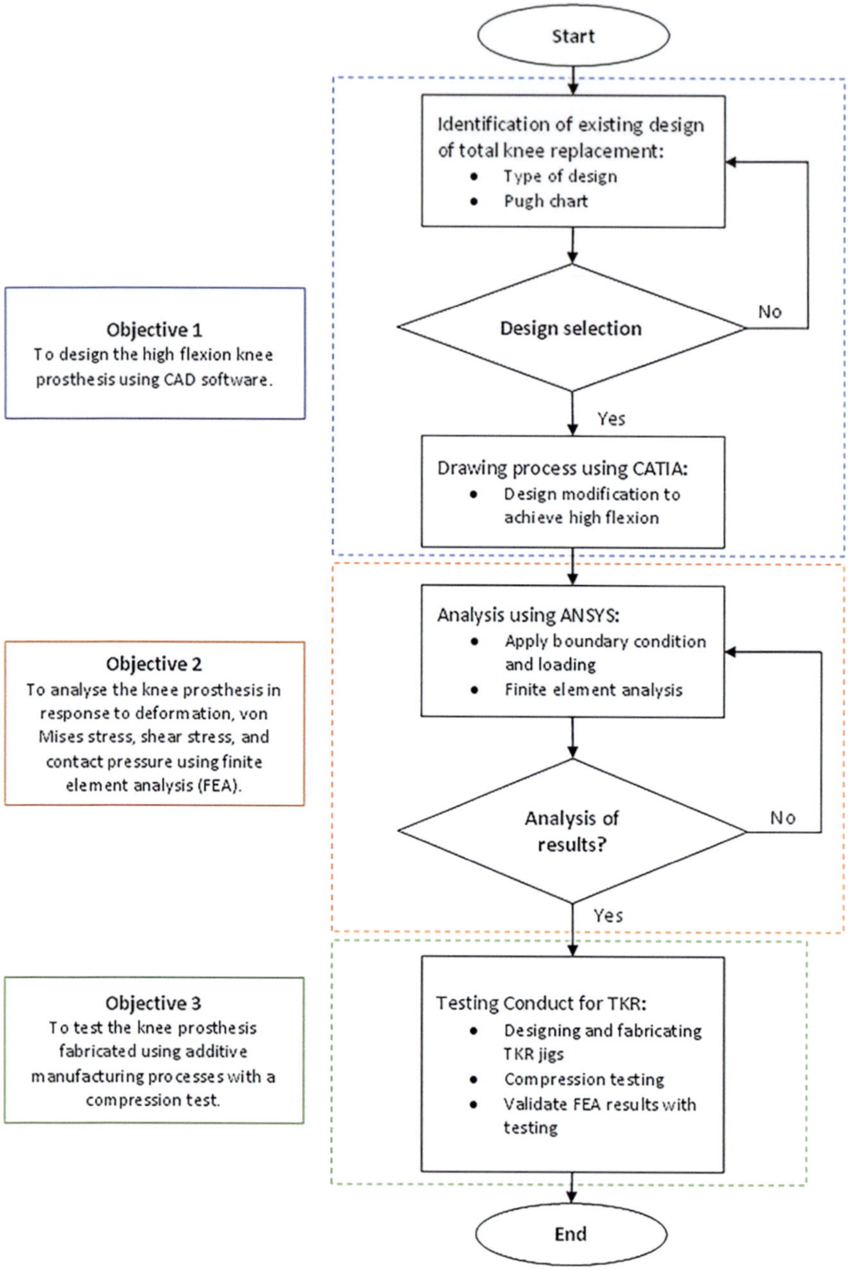

Fig. 4.1 Research flowchart

Table 4.1 Pugh chart of TKR design

Type of design			Cruciate retaining	Posterior stabilised	Mobile bearing
Image			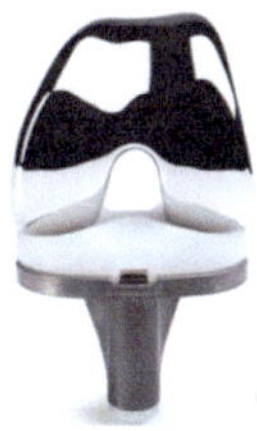	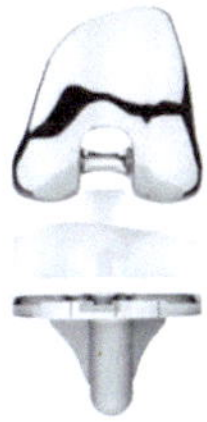	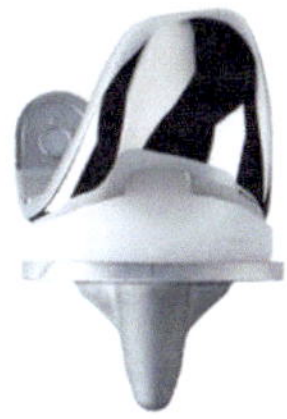
Criteria	Weight	Datum	Concept 1	Concept 2	Concept 3
Range of motion	3	0	+	+	+
Stability	3	0	+	+	+
Load transfer	2	0	+	+	+
Wear resistance	2	0	–	–	0
Easiness of surgery	1	0	–	+	+
Risk of surgery	1	0	0	0	0
+			8	9	7
0			1	1	3
–			3	2	2
Total			5	⑦	5

the Pugh chart to allow a wider range of motion. The datum was based on the NexGen LPS-Flex high-flexion total knee prosthesis from Zimmer Biomet, as shown in Fig. 3.2, which has the features for high-flexion TKR and can bend up to 155° (Fig. 4.2).

In collecting the current TKR designs, the cruciate retaining and posterior stabilised types were included in the Pugh chart. Both types are FB knee implants, so an MB knee implant was added to the Pugh chart for more comprehensive options. Furthermore, the Pugh chart required some necessary design criteria to compare with these selected types of TKR designs. Hence, the criteria for the selection of design were based on the range of motion, stability, wear resistance, load transfer, and more effortless surgical procedures. The weightage of each criterion was measured according to how significant each standard was in providing more range of flexion motion.

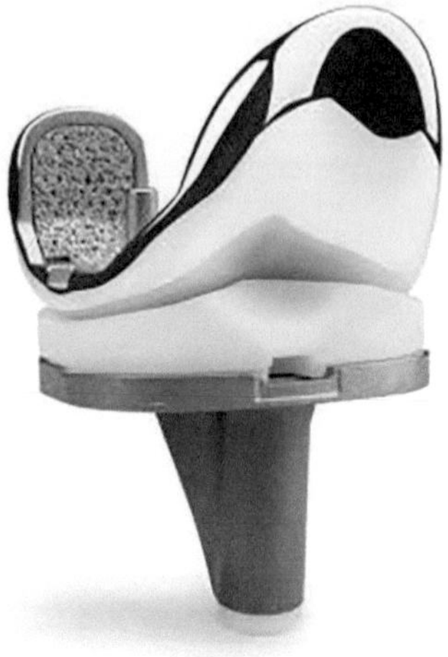

Fig. 4.2 NexGen LPS-Flex from Zimmer Biomet [1]

4.3 Design Modifications to Achieve High Flexion

The drawing began after selecting the design for TKR using computer-aided design (CAD) software. In the design modification process, all the features studied to provide a range of motion and stability to the TKR were applied to each component. After thoroughly researching and analysing each component development, posterior stabilised parts were drawn in using CAD software. The design development referred to NexGen LPS-Flex high-flexion total knee prosthesis from Zimmer Biomet for femoral, tibial, and tibial insert parts, as shown in Table 4.2.

The dimensions of femoral, tibial, and tibial insert components can be found in Figs. 4.3, 4.4, and 4.5, respectively. The detailed dimensions of the femoral, tibial, and tibial insert components are shown in Appendix 1, Appendix 2, and Appendix 3, respectively.

The modifications focused on the surface of the tibial insert as it has more impact on the contact area between the femoral and tibial insert components for increasing the range of motion. The actual design of the tibial insert was assigned in Fig. 4.5 and labelled as Design 1, as it was drawn from the beginning by going through the drawing process. Design 1 has the maximum thickness of 11 mm and a minimum thickness of 10 mm. The surface of the tibial insert was slightly curved. The modification started with Design 2, which has the maximum thickness of 11.2 and a minimum thickness of 8.6 mm to create an inclination surface on the tibial insert, as shown in Fig. 3.6. Meanwhile, Design 3 has a thickness of 10 mm with a flat surface on the tibial insert, as shown in Fig. 4.7. Table 4.3 shows all three different designs of tibial inserts. The detail dimensions of the tibial insert for Design 2 and Design 3 are shown in Appendix 4 and Appendix 5, respectively (Fig. 4.6).

Table 4.2 Components of total knee replacement

Part	Side view	Isometric view
Femoral	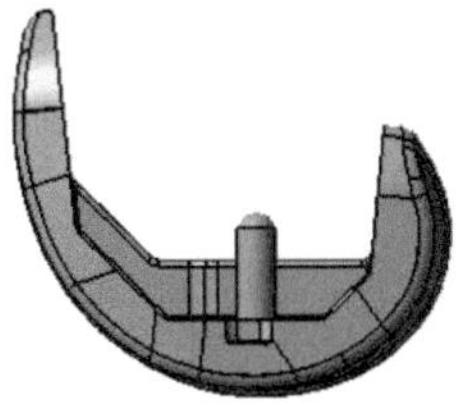	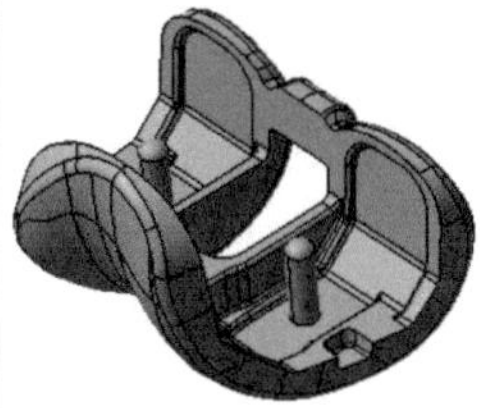
Tibial Insert	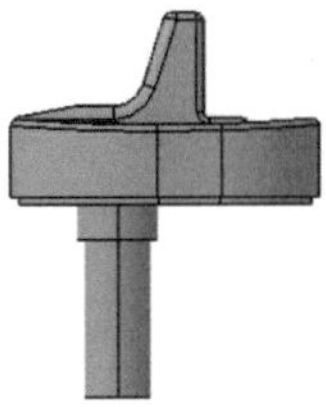	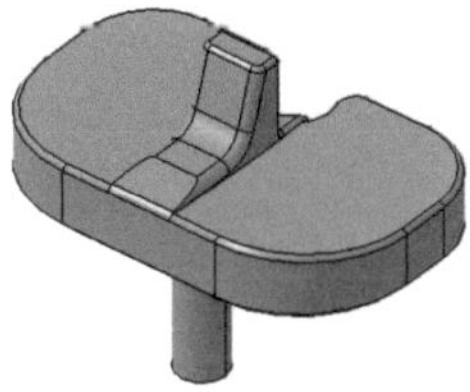
Tibial	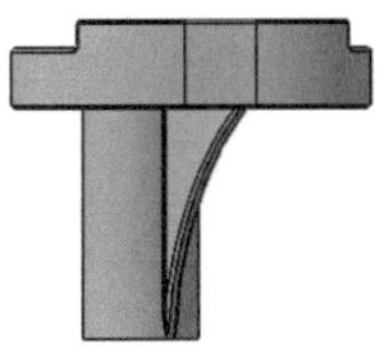	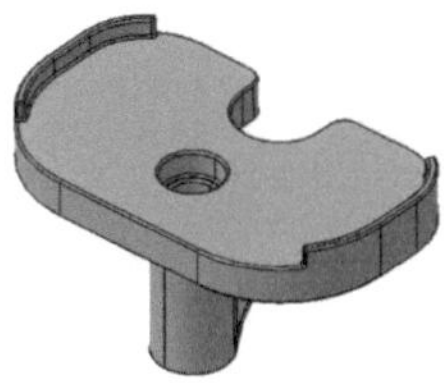

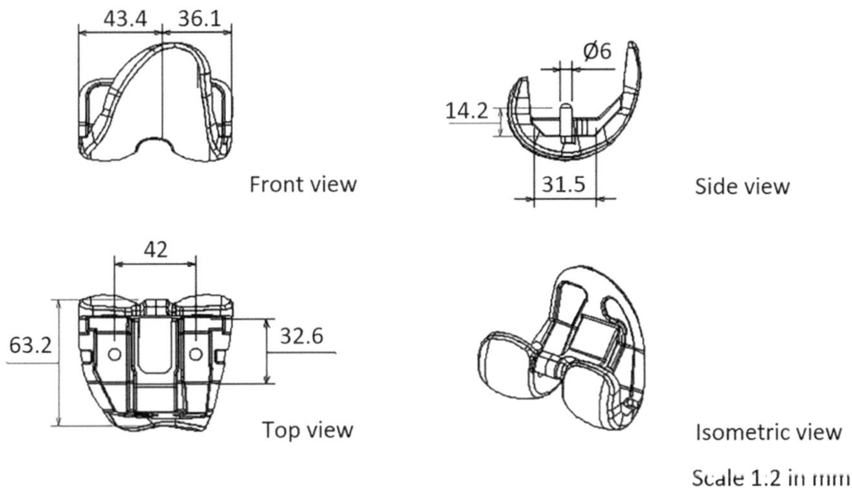

Fig. 4.3 Dimension of femoral component

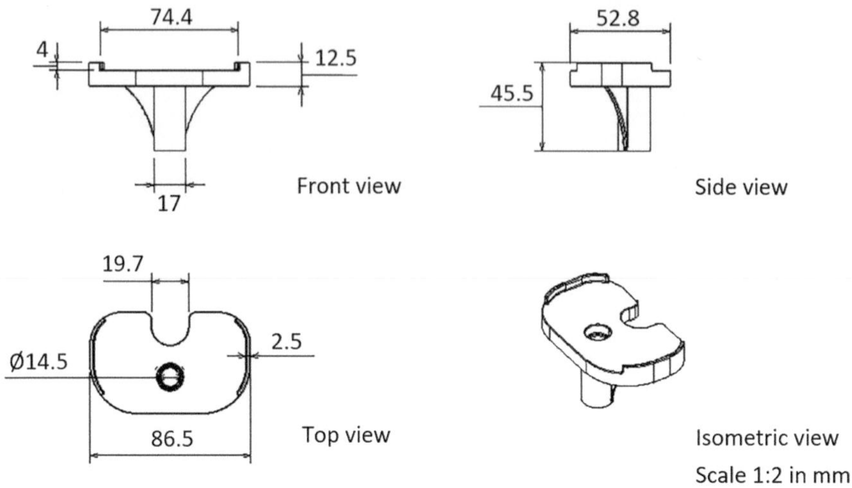

Fig. 4.4 Dimension of tibial component

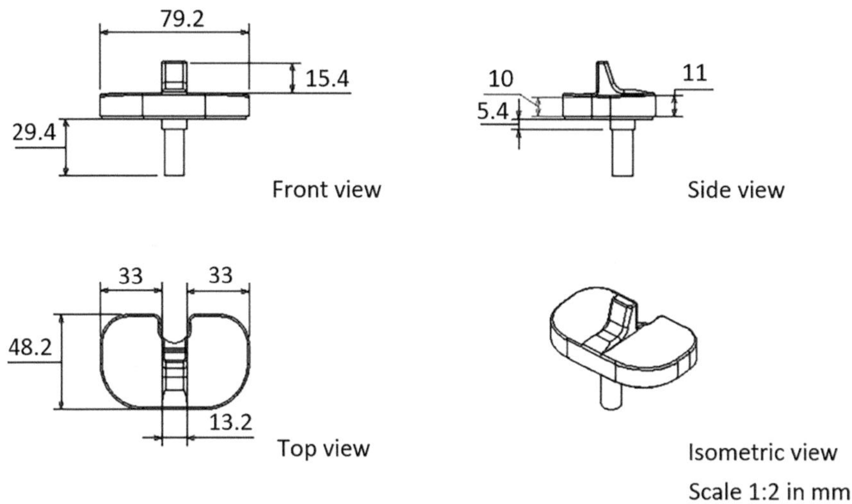

Fig. 4.5 Dimension of tibial insert (Design 1)

Then, the assembly of three different designs of the tibial insert with the femoral and tibial components was completed using CAD software, as shown in Table 4.4. The detailed dimensions of the assembly TKR components are shown in Appendix 6.

Table 4.3 Different designs of tibial insert

No. of part	Side view	Isometric view
Design 1	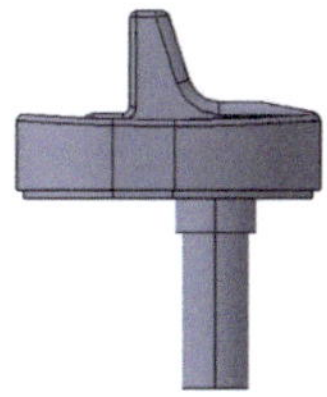	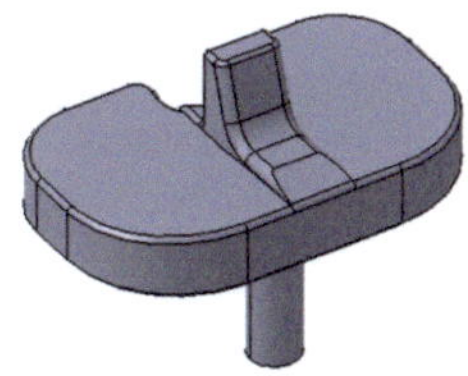
Design 2	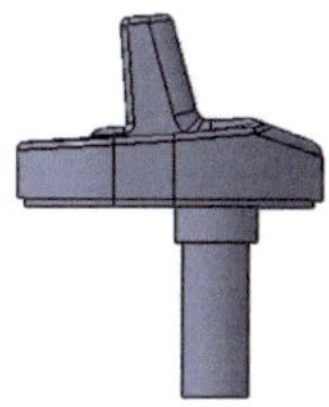	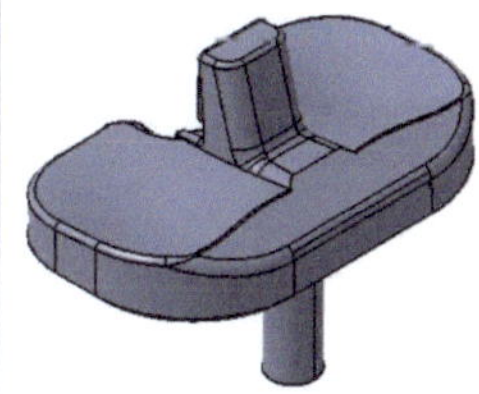
Design 3	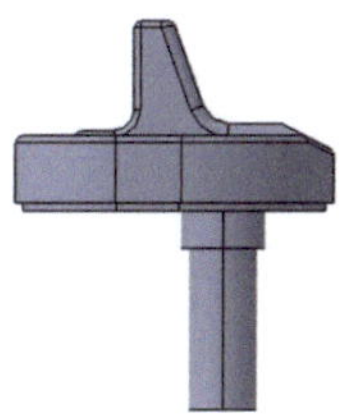	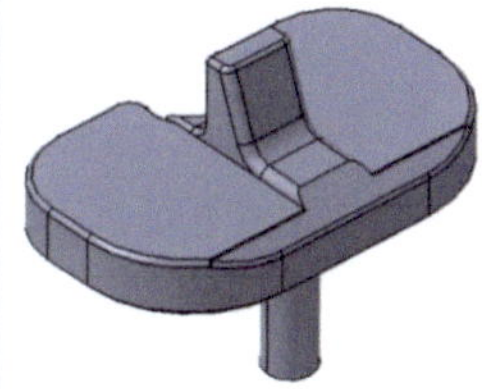

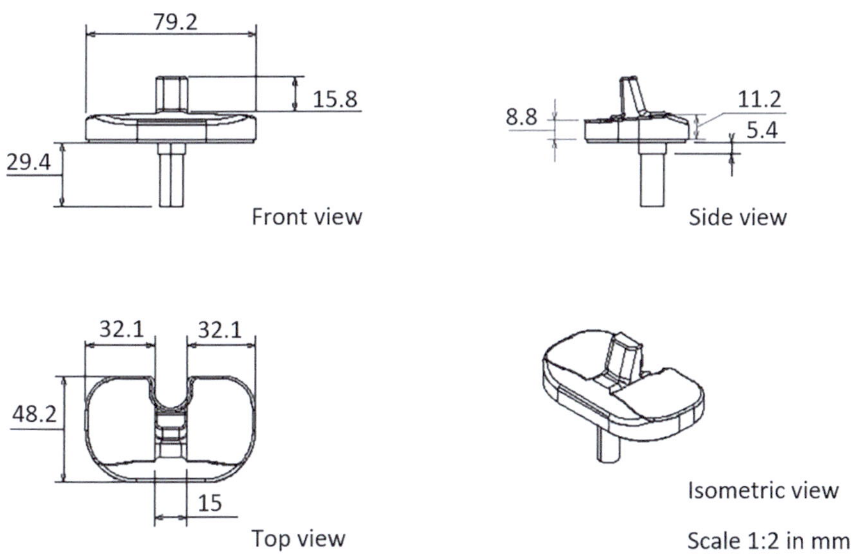

Fig. 4.6 Dimension of tibial insert (Design 2)

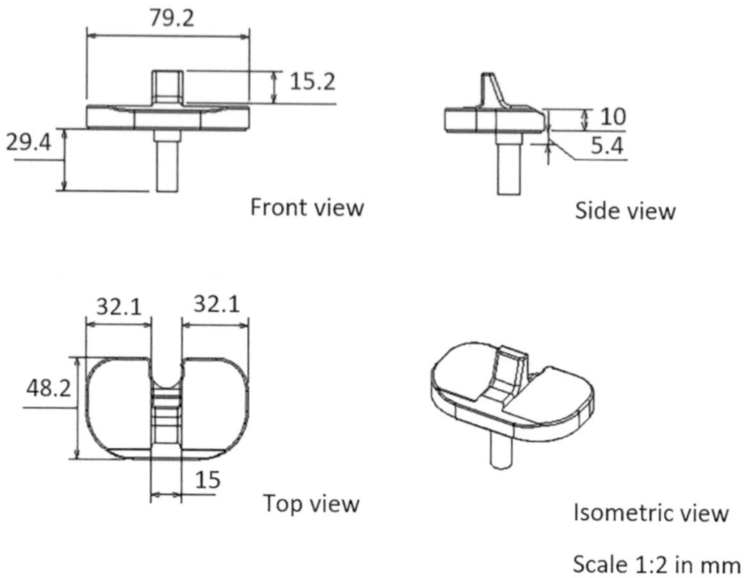

Fig. 4.7 Dimension of tibial insert (Design 3)

4.4 Finite Element Analysis (FEA) on TKR

After completing the component assembly, the following process was to conduct finite element analysis using CAE software. Additionally, the FEA was carried out to observe the performance of three different designs of the tibial insert along with the improvements of the femoral and tibial components. Firstly, the assembled parts file from the CATIA software must change into a suitable format that the ANSYS software can read. For instance, the CAT file was saved in IGES format before import to the ANSYS software. Then, the material properties assigned to the femoral, tibial insert, and tibial components, as shown in Table 4.5, were assumed to be homogeneous and isotropic. Young's modulus and Poisson ratio of titanium alloy and polyamide were determined based on the reported values in prior studies [2, 3].

A contact surface between the different TKR components was determined in the simulation. The contact region consists of contact bodies assigned to the tibial insert component and target bodies assigned to the femoral and tibial components. Hence, the contacts between the femoral and tibial insert components and the tibial insert and tibial components were modelled in the simulation, resulting in two contact pairs, as shown in Table 4.6. In addition, the interaction between contact and target bodies was modelled as fully bonded.

The assembly components were subjected to a finite element mesh with an adaptive tetrahedral element using SOLID187, CONTA174, and SURF154 type elements. A finite element size of 1.9 mm and 2.0 mm was applied in the meshing process, as shown in Fig. 4.8. Furthermore, to evaluate the accuracy of the computational result,

Table 4.4 Assembly of all TKR components

No. of part	Side view	Isometric view
Design 1	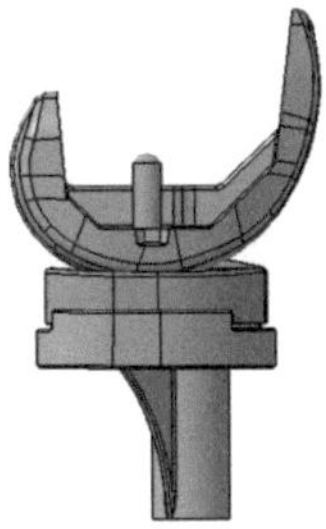	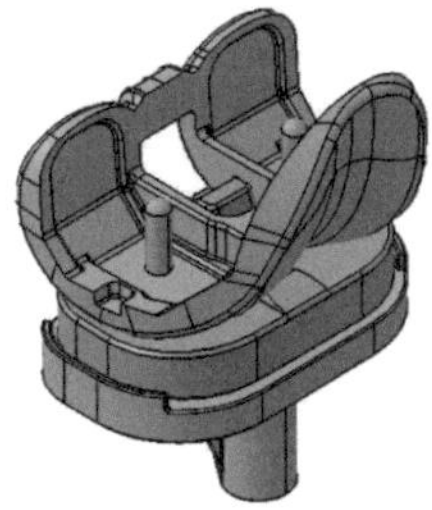
Design 2	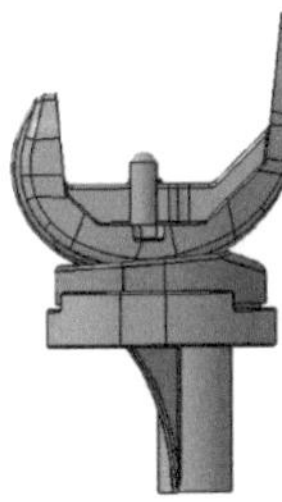	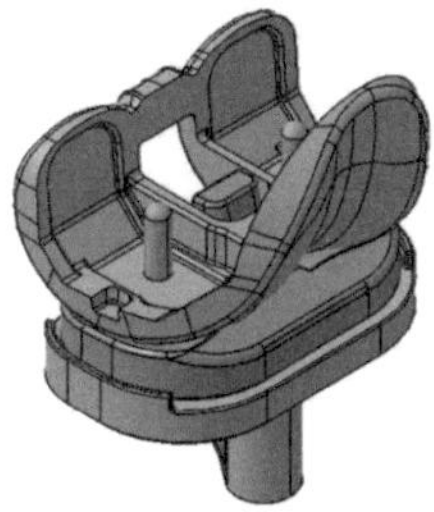
Design 3	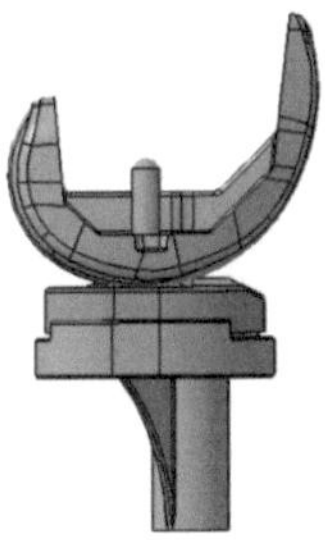	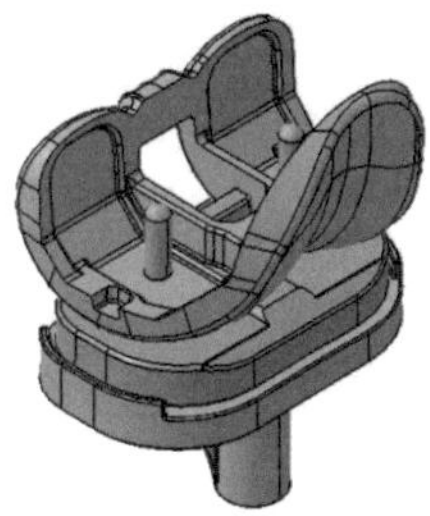

Table 4.5 Materials properties of components in TKR

Component	Material	Poisson ratio	Young's modulus (MPa)
Femoral and Tibial	Titanium alloy	0.3	110,000
Tibial insert	Polyamide	0.394	1,701

a convergence study of the mesh utilising adaptive elements with sizes ranging from 1 to 10 mm was carried out. After acquiring the optimised mesh, the finite element deformations and stress analysis were ready to be used. Table 4.7 presents the number of nodes and elements for TKR components during the meshing process.

Then, boundary conditions such as fixed support were assigned to the base of the tibial part to display as cemented TKR, as shown in Fig. 4.9. Four different

Table 4.6 Contact region of the assembly TKR

No. of contact	Front view	Isometric view
Contact 1	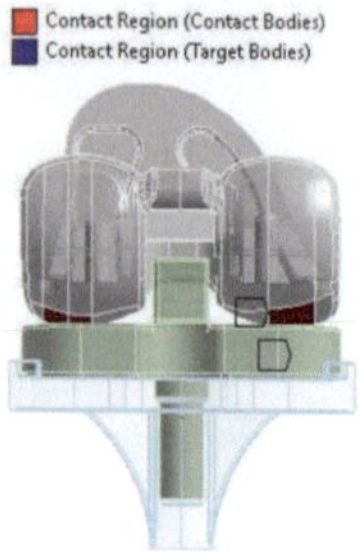	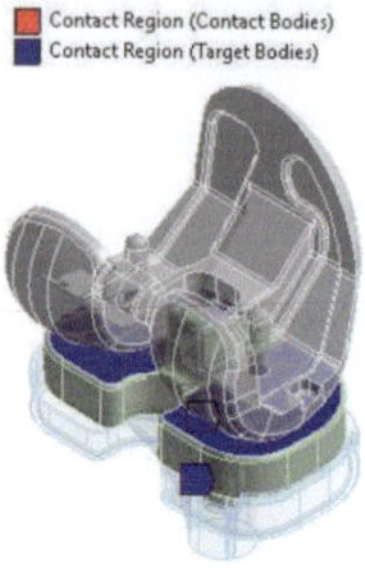
Contact 2	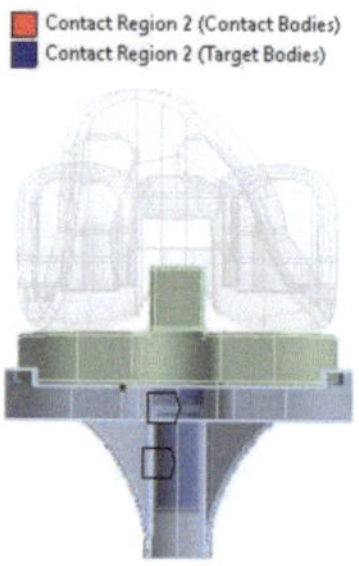	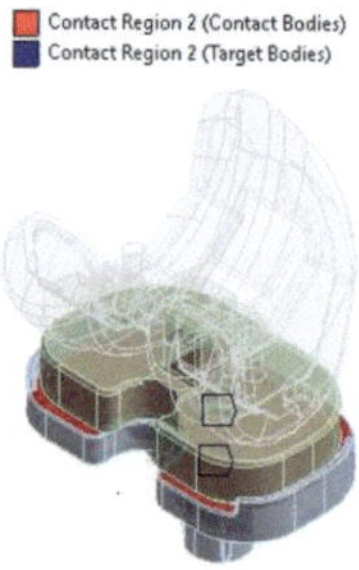

angles were chosen based on the most common daily activities of Asians. Before analysis begins, angles of flexion were introduced on the assembled parts, such as 0°, 90°, 135°, and 165° that represent standing, kneeling, squatting, and sitting on feet, respectively, as shown in Table 4.8.

The use of net forces depends on the percentage of body weight to recreate knee load during standing, kneeling, squatting, and sitting on feet. These net forces were calculated from the percentage of body weight times the average body weight. The average body weight applied was 652.365 N, equivalent to 66.5 kg applied vertically downward towards the Z-axis. The various net forces were applied to each position depending on the flexion angle, as shown in Table 4.9.

The outcome of FEA includes total deformation, von Mises stress, and contact pressure of the assembly TKR. Total deformation measured the deformation of TKR components under the load, and von Mises stress predicted the materials' yields under the force applied in the Y-direction. Meanwhile, contact pressure measured the reaction to the load applied to the contact area. In this study, the contact area consists of the femoral component with the tibial insert and the tibial insert with the tibial component.

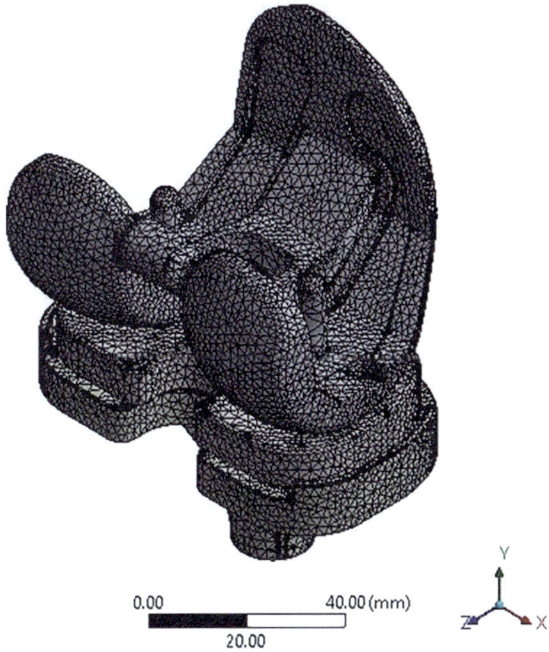

Fig. 4.8 Isometric view of assembled parts for the meshing process

Table 4.7 Number of nodes and elements for all TKR components

Number of nodes	Number of elements
427,391	291,041

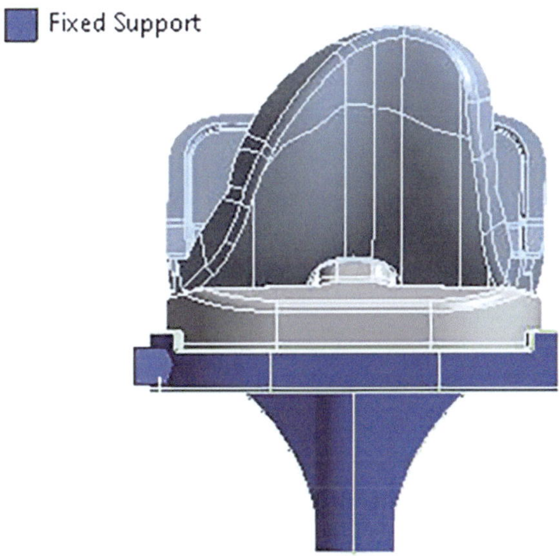

Fig. 4.9 Fixed support on the tibial component (front view)

Table 4.8 Angle of flexion

Angle of flexion	Assemble part	Position
0°	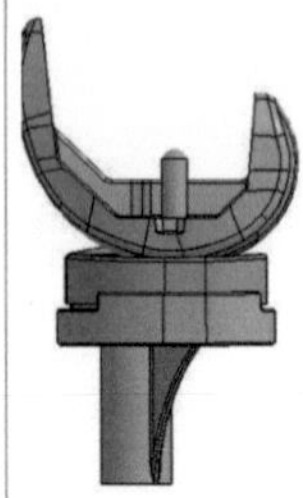	Standing
90°	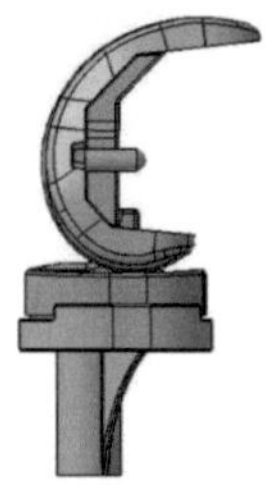	Kneeling
135°	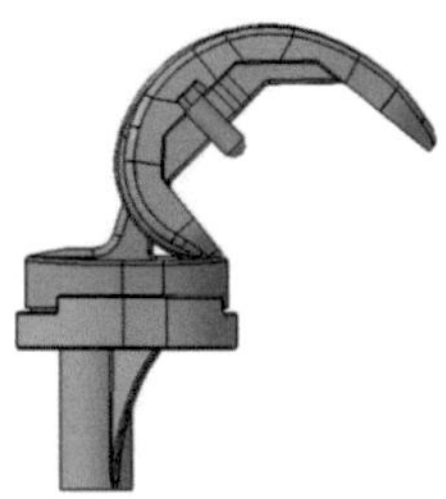	Squatting
165°	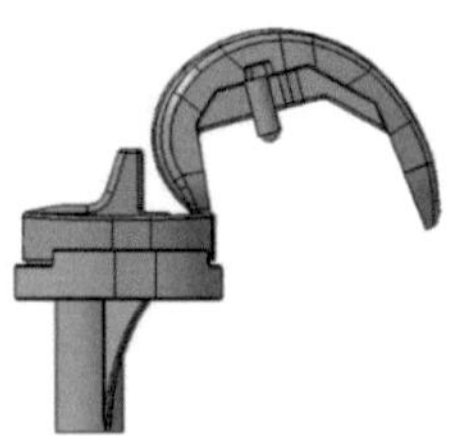	Sit-on-Feet

4.5 Preparation of Jigs and Compression Testing on TKR

The finalised modified TKR was fabricated following the simulation using additive manufacturing (AM) technology. The fabrication process was completed using 3D printing to preserve time, as the TKR components were considered customised with complex geometry. The selective laser sintering (SLS) process was assigned

Table 4.9 Force applied for each position

Angle of flexion	Position	Force applied
0°	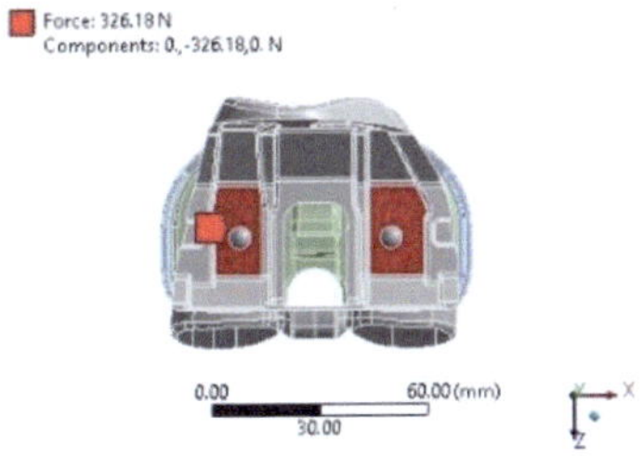	Force = 0.5 × 652.365 N = 326.182 N
90°	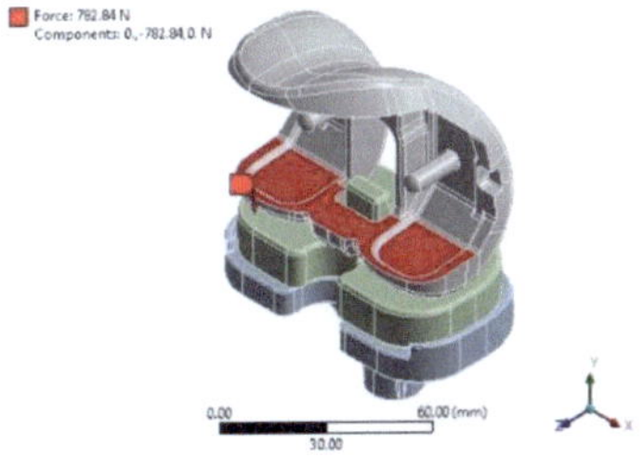	Force = 1.2 × 652.365 N = 782.838 N
135°	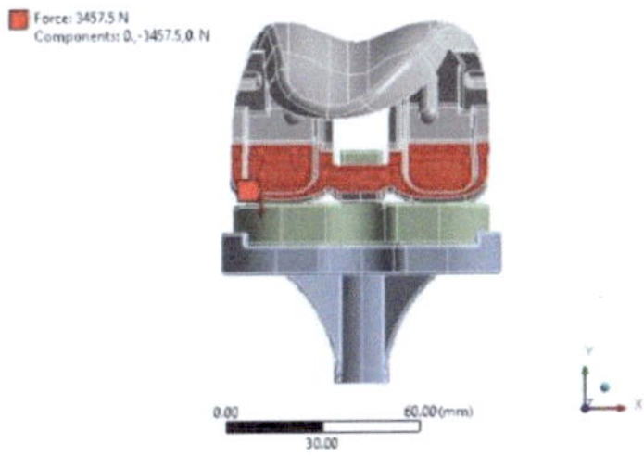	Force = 5.3 × 652.365 N = 3457.535 N
165°	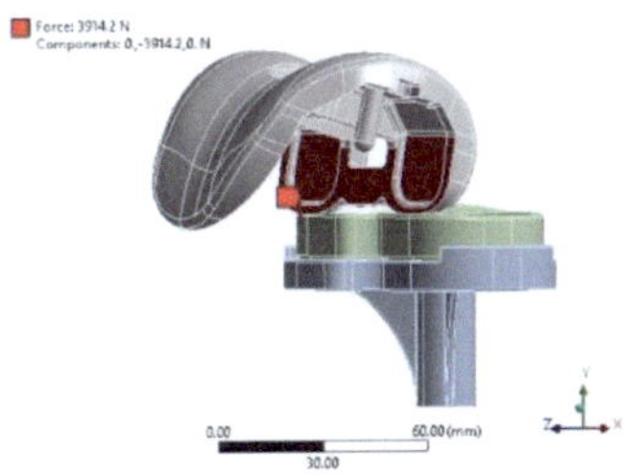	Force = 6.0 × 652.365 N = 3914.190 N

to print the femoral and tibial components with titanium alloy material using the RENISHAW RenAM 500E Metal 3D printer, as shown in Plate 4.1. Before and after finishing process were also included, as shown in Plates 4.2 and 4.3, respectively. Meanwhile, the fused deposition modelling (FDM) process was assigned to print the tibial insert component with polyamide material using 3DGence Industry F350 printer. The complete fabricated tibial insert can be seen in Plate 4.4. All these fabrication processes were accomplished by outsourcing to the 3D Gens Sdn Bhd, a 3D printing company.

The assembly component testing was conducted after all the TKR components had been printed completely. Before testing for the components, a jig was prepared to accommodate the TKR components to fit well around the test machine. Two grips from the upper side and lower side of the machine could perfectly hold the femoral and tibial components. Hence, two jigs were designed using CAD software to accommodate the femoral and tibial components throughout the testing. The jig

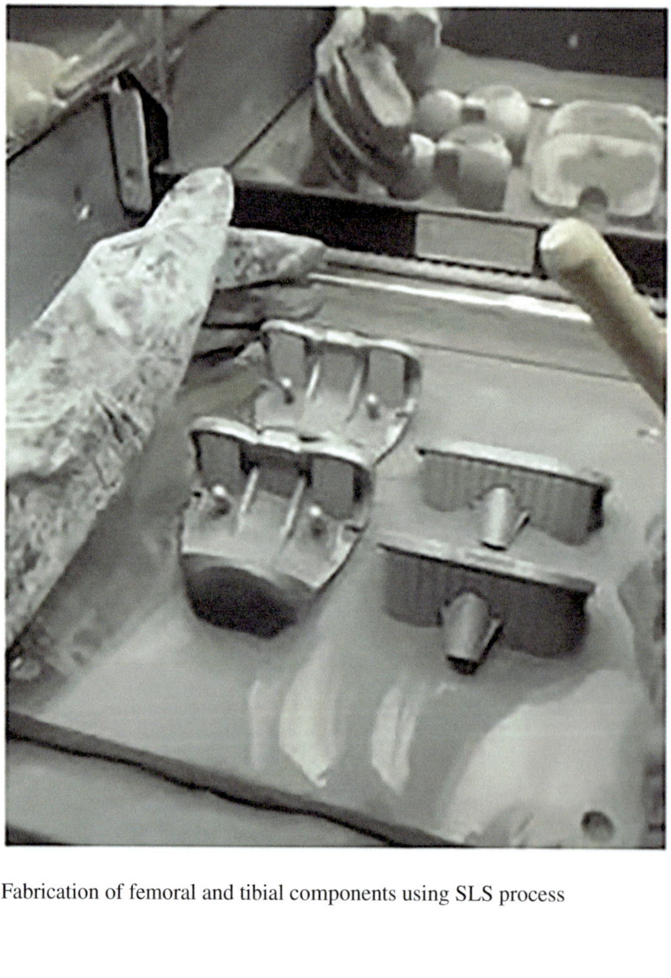

Plate 4.1 Fabrication of femoral and tibial components using SLS process

Plate 4.2 Before finishing the femoral and tibial components

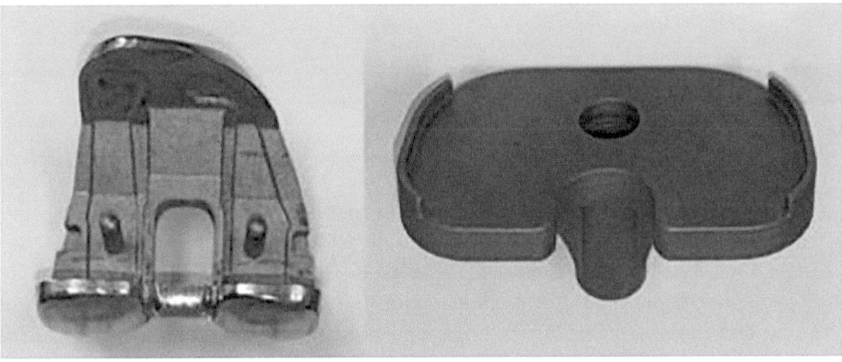

Plate 4.3 After finishing the femoral and tibial components

preparation began by drawing the femoral jig with a shaft to fit between the upper side grip and two holes on the bottom of the femoral jig to insert inside the femoral component, as shown in Table 4.10. The dimension of the femoral jig component can be seen in Fig. 4.10. Therefore, the holes on the bottom of the femoral jig component were inserted entirely into the femoral component, and the shaft of the femoral jig

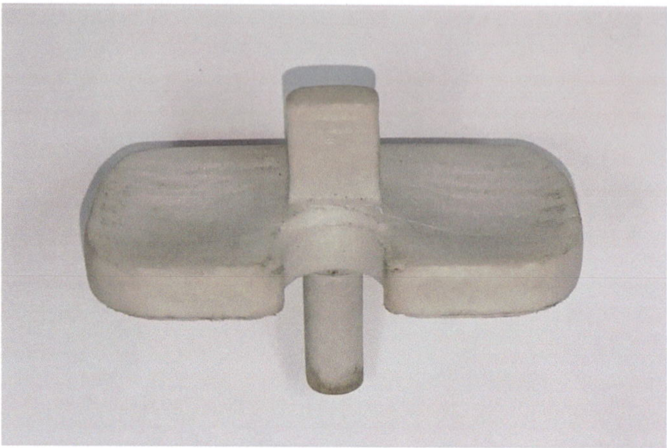

Plate 4.4 Fabricated tibial insert component

Table 4.10 Femoral jig component

Front view	Bottom view	Isometric view
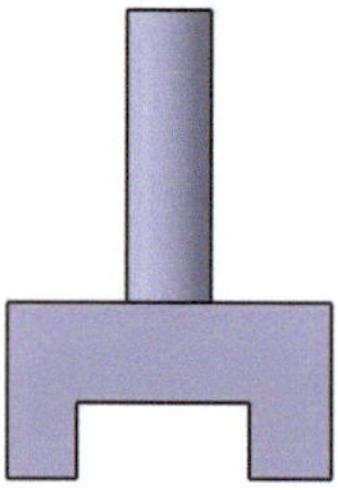		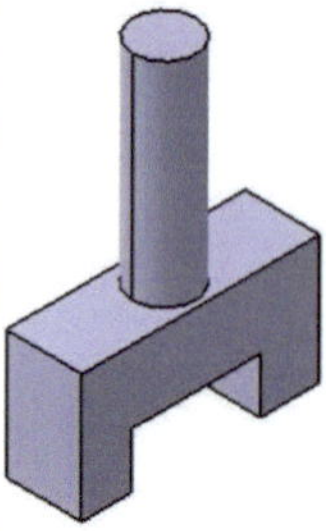

component adjusted well to the testing machine, as shown in Table 4.11. The detailed dimensions of the femoral jig component as shown in Appendix 7.

Then, the tibial jig was drawn with a shape of tibial circumference on top of the tibial jig to insert the tibial component entirely. A hole thread in the bottom of the tibial jig was drawn for a shaft to hold the tibial jig, as shown in Table 4.12. The dimension of the tibial jig component can be found in Fig. 4.11. Therefore, the top of the tibial jig was cut off in the shape of the tibial component to place the tibial insert and tibial components together, so a shaft to hold the tibial jig was adjusted to fit in the testing machine, as shown in Table 4.13. The detailed dimensions of the tibial jig and holder for tibial jig components are shown in Appendix 8 and Appendix 9, respectively.

Afterwards, the femoral and tibial jigs were fabricated with stainless steel material and completed with coatings to prevent rusting, as shown in Plates 4.5 and 3.6, respectively. Stainless steel was chosen because it has Young's modulus of 190 GPa

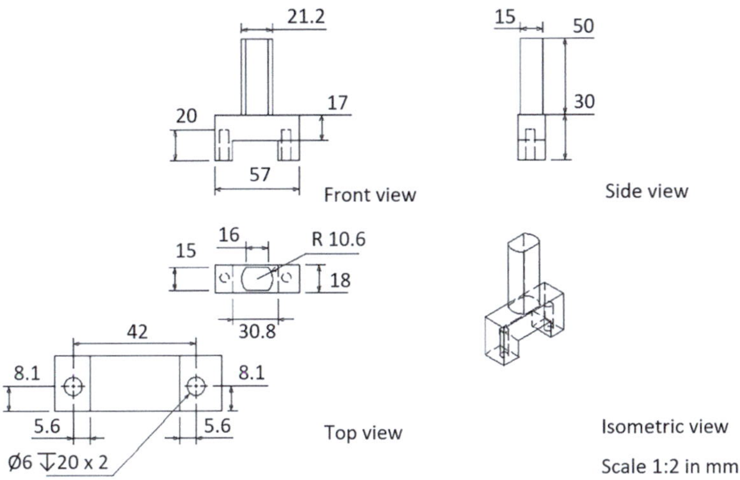

Fig. 4.10 Dimension of the femoral jig

Table 4.11 Assembly of femoral jig and femoral components

Front view	Side view
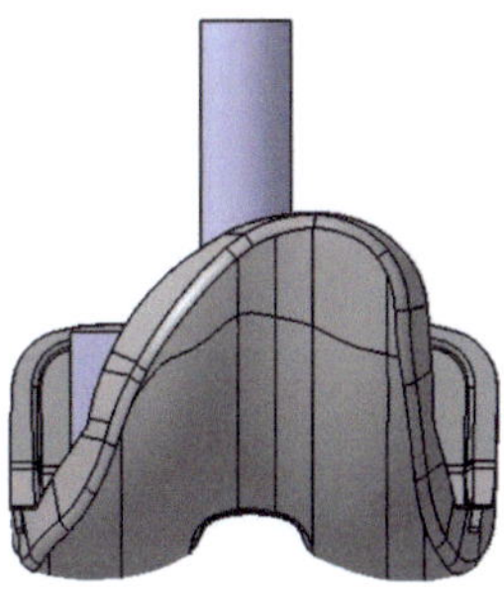	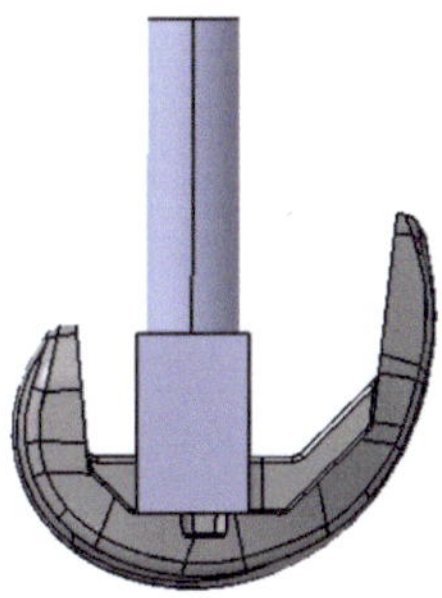

Table 4.12 Tibial jig component

Top view	Bottom view	Isometric view
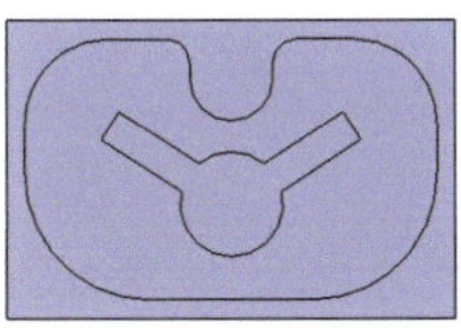	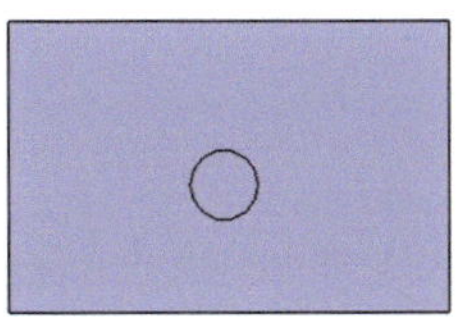	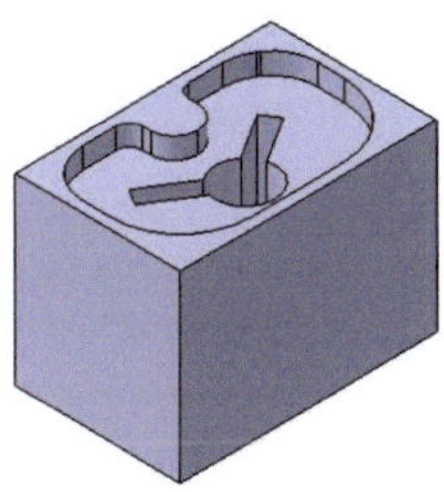

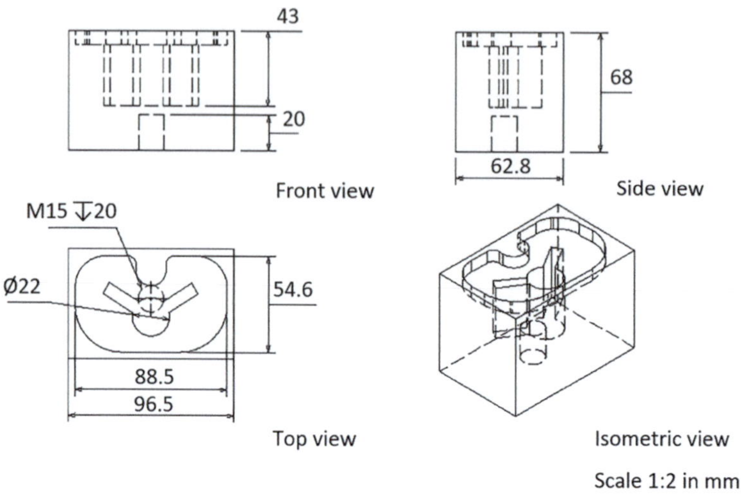

Fig. 4.11 Dimension of the femoral jig

Table 4.13 Assembly of tibial jig, tibial insert, and tibial components

Front view	Side view
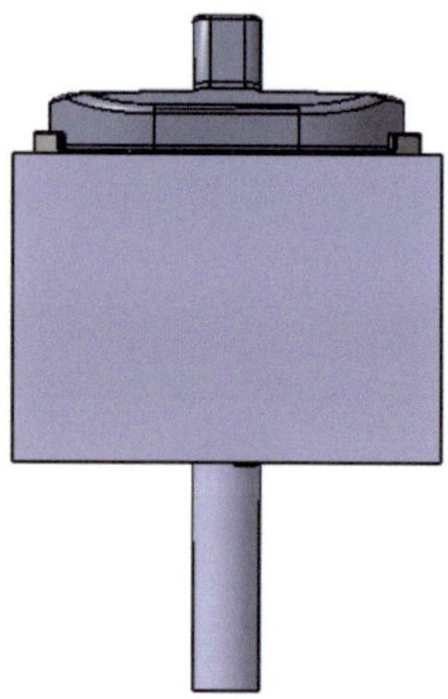	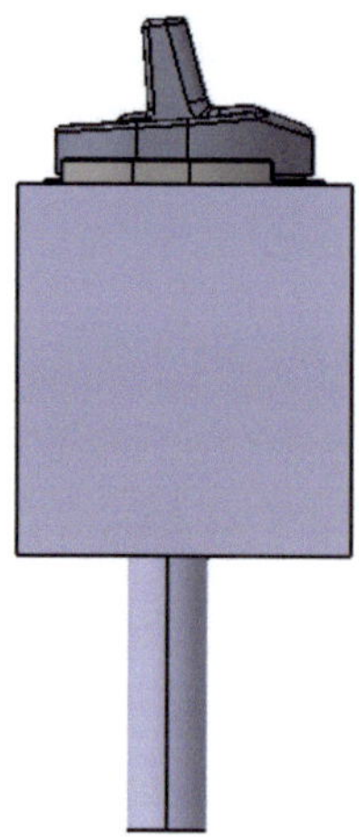

to withstand deformation. Besides, it has more strength than titanium alloys, so it can effortlessly transfer force to the femoral component (Plate 4.6).

A compression test can now be conducted with the arrangement starting from the top of the test machine as the femoral jig, followed by the femoral component, tibial insert component, tibial component, and, lastly, tibial jig. The compression test arrangement can be referred to as shown in Plate 3.7. The detailed dimensions of the

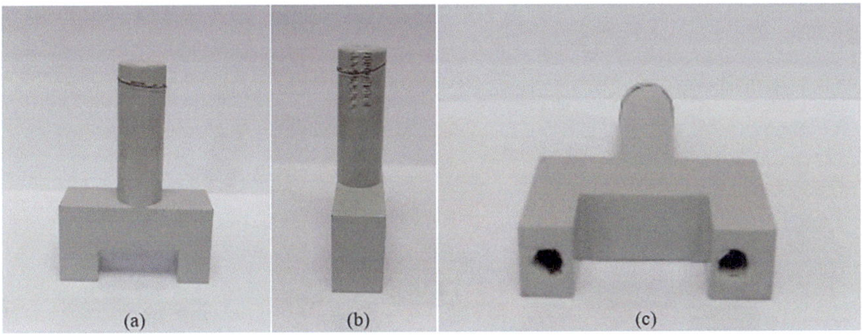

Plate 4.5 **a** Front view, **b** Side view, and **c** Bottom view of the femoral jig

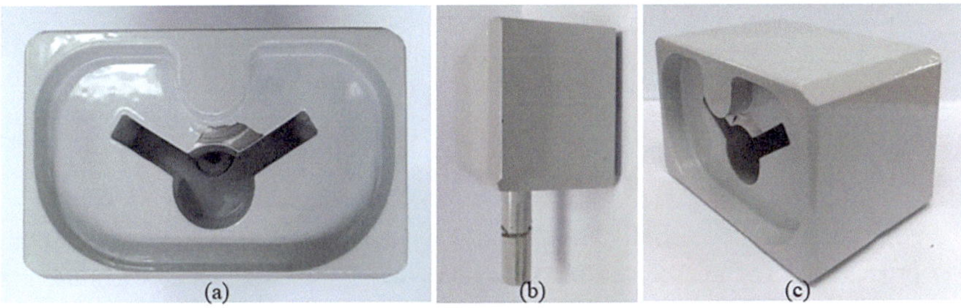

Plate 4.6 **a** Top view, **b** Side view, and **c** Isometric view of the tibial jig

assembly for jigs and TKR components are shown in Appendix 10 and Appendix 11, respectively (Plate 4.7).

After all the components involved in the testing were appropriately set in the Shimadzu Servopulser testing machine, the force displacement of the actuator was applied at a speed of 0.5 mm/min to load on the femoral. The load was applied until it reached 326.18 N, representing the standing position at the angle flexion of 0°, as stated in Sect. 3.3 and Table 3.9, which took about 20 to 30 s. The 4890 Controller and GLUON software were used to measure and collect the displacement and force data, respectively. The displacement or deformation was measured on the assembly of TKR generated by loading applied vertically to the femoral component. In addition, the compression testing referred to the standard of ASTM F2777 for evaluating tibial insert endurance and deformation under high flexion, which neglected the endurance section. The details for the standard of ASTM F2777 are shown in Appendix 12.

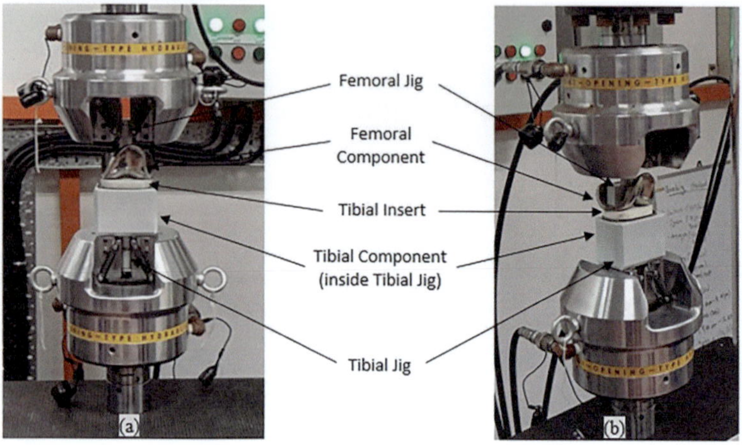

Plate 4.7 **a** Front view and **b** Isometric view of the compression test

4.6 Summary

The Pugh chart has been briefly introduced in the design selection. The design selection process also addressed the modifications to increase the range of motion or flexion for the TKR component. The drawing process of the TKR parts began by using CAD software. Loading and boundary conditions were successfully applied to the modified design in the finite element analysis using software. The process of fabrication using AM for knee prosthesis was stated briefly. The preparations of jigs for femoral and tibial components were utterly elaborated. Therefore, the next chapter will discuss the design selection process, the outcomes of FEA in terms of total deformation, von Mises stress, and contact pressure, and the results of compression testing for fabricated TKR.

References

1. NexGen® Complete Knee Solution | Proven, "Supports High-Flexion Activities," Zimmer Biomet. [Online]. Available: https://www.zimmerbiomet.com/en/products-and-solutions/specialties/knee/nexgen-complete-knee-solution.html. Accessed 31 July 2021
2. M. Long, H.J. Rack, Titanium alloys in total joint replacement-a materials science perspective. Biomaterials **19**(18), 1621–1639 (1998)
3. D.A. Porter, M.A. D. Prima, Y. Badhe, A.R. Parikh, Nylon lattice design parameter effects on additively manufactured structural performance. J. Mech. Behav. Biomed. Mater. **125** (2022)

Chapter 5
Finite Element Analysis (FEA) and Experimental

5.1 Introduction

The results for each implemented method to develop high-flexion total knee replacement are presented in this chapter. The outcomes were evaluated and researched sufficiently to understand the concept and resolve the research questions. As stated in the preceding chapter, the modifications of the tibial insert components were divided into three designs capable of increasing the angle flexion of the knee prosthesis. During the FEA, the modified tibial inserts were introduced with different angle flexions, such as 0°, 90°, 135°, and 165° that represent standing, kneeling, squatting, and sitting on feet, respectively. Static structural analysis was implemented to analyse the modified assembly of TKR under loadings and boundary conditions at different angle flexions.

Subsequently, the results of FEA were analysed and compared in total deformation (mm), von Mises stress (MPa), and contact pressure (MPa) of the assembly TKR. Hence, the main finding in the FEA was to find the optimum design of modified TKR at 0°, 90°, 135°, and 165° angle flexions. Then, the finding of the optimum modified design of TKR was fabricated using additive manufacturing processes. Suitable jigs for optimum modified TKR were designed and manufactured to fit in the testing machine during the compression testing. The FEA and compression test results were compared and discussed thoroughly.

5.2 Knee Prosthesis Design

During the design selection process, the Pugh chart was applied to select the type of knee prosthesis designs categorised into three design types: posterior stabilised, cruciate retaining, and mobile bearing designs. These design types were assigned to the Pugh chart because of the ability to increase the range of motion in a knee

I. M. Arrif et al., *High-Flexion Knee Prosthesis*,
SpringerBriefs in Applied Sciences and Technology,
https://doi.org/10.1007/978-981-96-9878-3_5

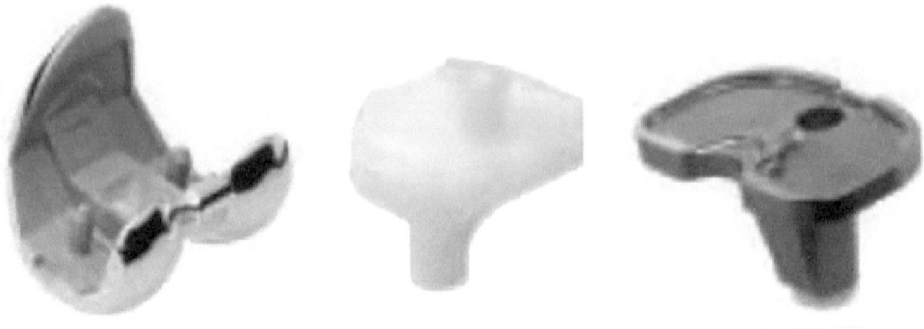

Fig. 5.1 Parts of the NexGen LPS-Flex Knee [1]

prosthesis. The design selection criteria include a range of motion, stability, wear resistance, load transfer, and simplify surgical procedures. Nevertheless, PS knee prosthesis can increase the range of motion, stability, load transfer, and more easiness in surgical procedures. Therefore, after considering all the criteria, the PS design was chosen by achieving the highest mark compared to the CR and MB designs.

Moreover, the PS design has a post-cam mechanism, which is a great feature compared to the CR design, to prohibit excessive femoral glide. By adding height extension to the cam located at the femoral component, the point loading between the femoral component and the tibial insert can be maintained and reduced during flexion angles up to 155°. An additional feature of the reference design includes the locking mechanism, which has an opening tube located at the tibial component and the stem protruding from the tibial insert. The stem of the tibial insert is inserted into the opening tube of the tibial component to allow more movement when using the TKR implant.

The reference design for high-flexion TKR was the NexGen LPS-Flex high-flexion total knee prosthesis from Zimmer Biomet, as shown in Figs. 5.1 and 5.2.

The development and design were created for femoral, tibial, and tibial insert parts using CAD software, as shown in Figs. 5.3, 5.4, and 5.5, respectively.

5.3 Modifications of High-Flexion TKR

Increasing the thickness of the tibial insert component would reduce contact pressure and increase the contact area. Besides that, the design modification also focused on connecting the femoral component and tibial insert curvature. The design of the tibial insert curvature determines the contact area between the femoral component and the tibial insert. Moreover, the tibial insert is one of the critical parts in modifying knee prosthesis because of contact pressure when bearing the load. Therefore, the connection between the femoral component and the tibial insert curvature, and thickness of

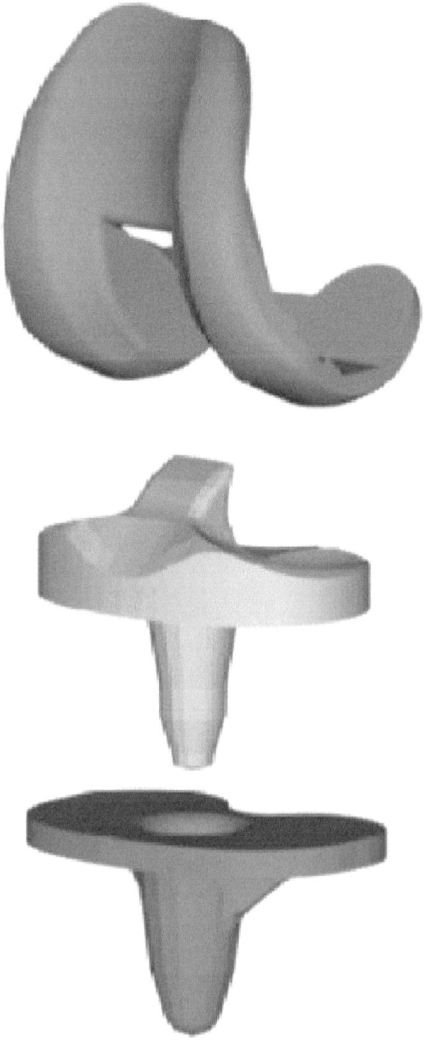

Fig. 5.2 Parts of the NexGen LPS-Flex Knee [2]

tibial insert are crucial for increasing the range of motion. Figures 5.6 and 5.7 show the modifications made for tibial insert parts.

Table 5.1 shows the modifications made for Design 2 and Design 3 to increase the range of motion and includes descriptions for Design 1. The cross-section of assembly for all the designs was presented to compare the contact area between the femoral component and the tibial insert curvature.

One feature of high flexion includes a wider tibial insert contact surface to lower wear and contact pressure. Therefore, the modified tibial insert must obtain stresses and pressure lower than the yielding stress of the material assigned to the tibial insert component. Increasing the flexion angle of TKR decreases the stresses and pressure

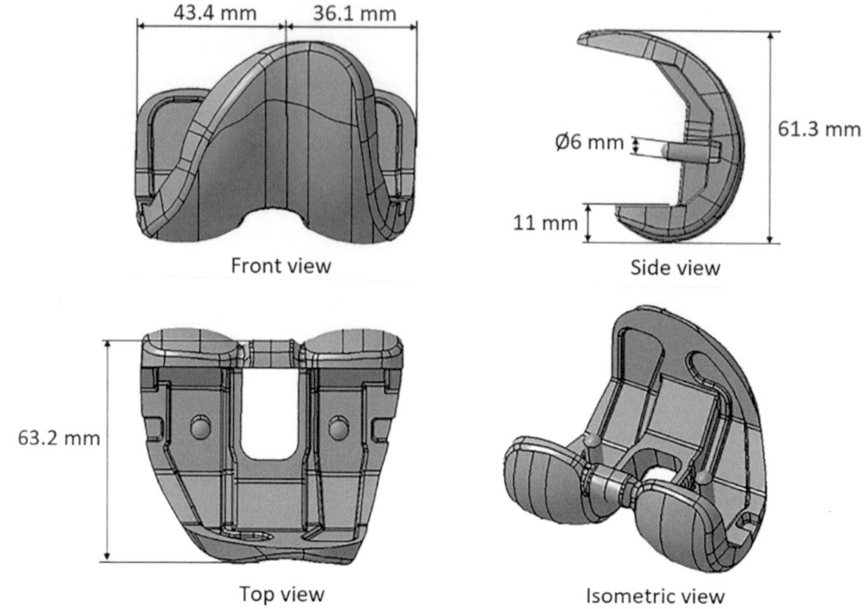

Fig. 5.3 Schematic diagram of Femoral Component

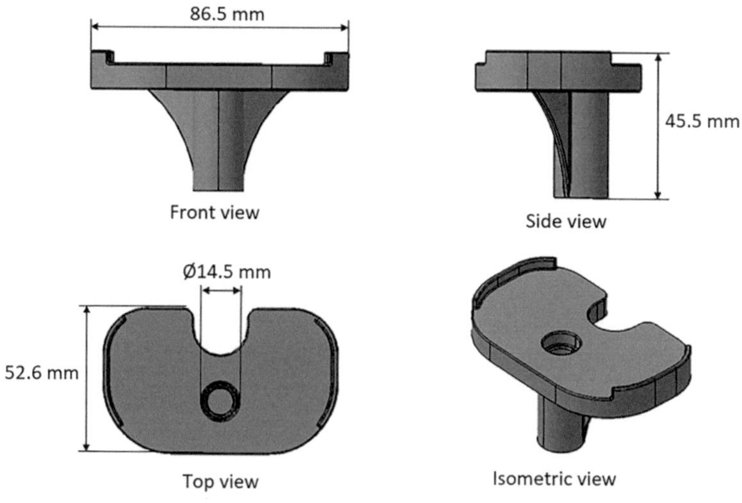

Fig. 5.4 Schematic diagram of Tibial Component

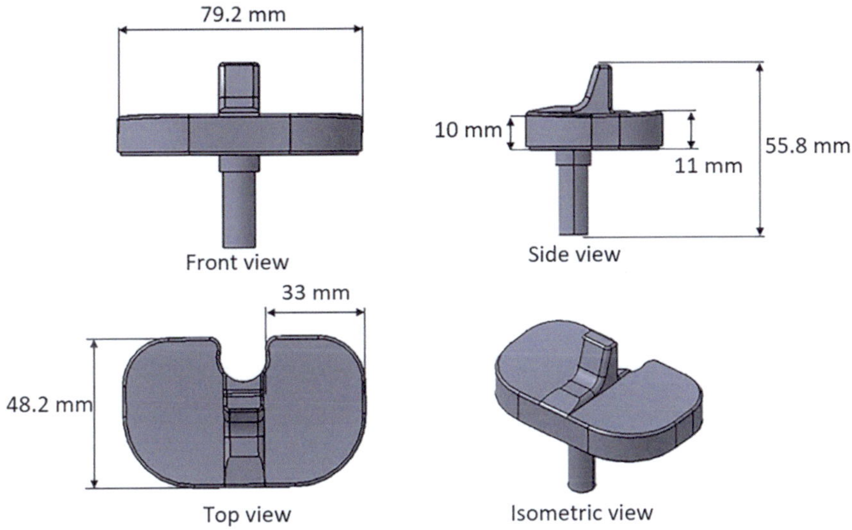

Fig. 5.5 Schematic diagram of Tibial Insert (Design 1)

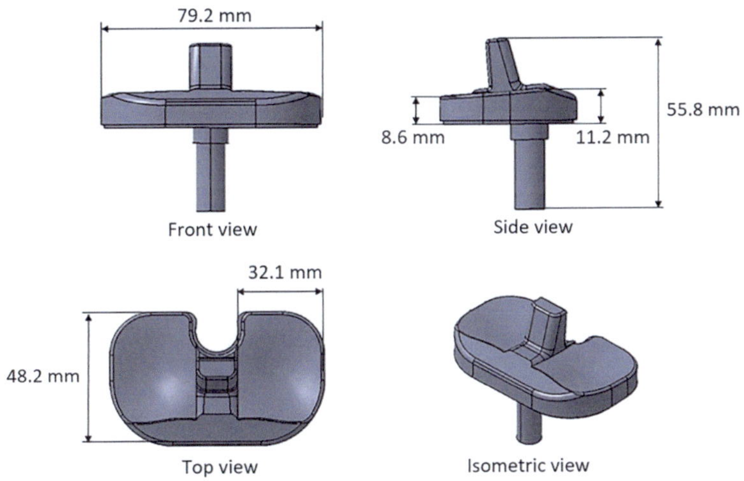

Fig. 5.6 Schematic diagram of Tibial Insert (Design 2)

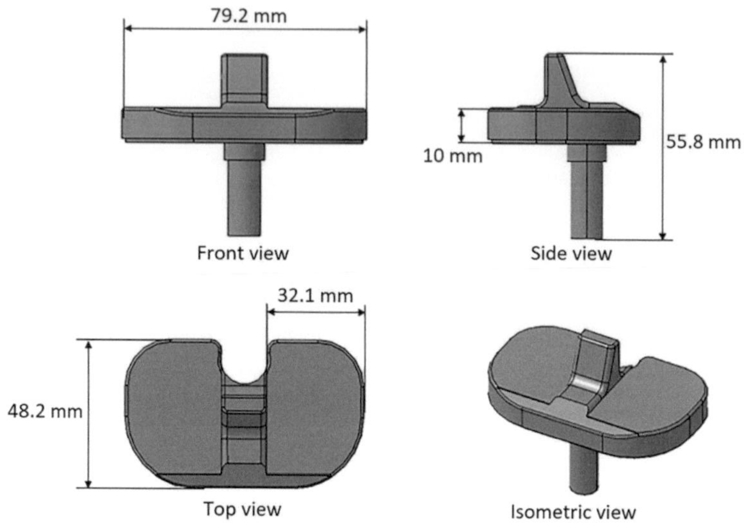

Fig. 5.7 Schematic diagram of Tibial Insert (Design 3)

involved in TKR to prevent wear and fracture. In conclusion, the modified TKR with three different contact areas between the femoral and tibial insert components shows potential for better knee flexion.

5.4 Finite Element Analysis (FEA) Results of Modified Assembly TKR

The FEA process applied static structural analysis to the modified assembly TKR. During the analysis, loads were applied to the femoral component to imitate the real knee joint of applying load on the femur bone. The angle of flexion used was 0°, 90°, 135°, and 165° to represent standing, kneeling, squatting, and sitting on feet, respectively. Hence, body weights were calculated to implement the different loads with four flexion angles. The analysis performed on the modified TKR obtained results on total deformation (mm), von Mises stress (MPa), and contact pressure (MPa).

The results display a colour code range from blue to red to present the behaviour of total deformation graphically, von Mises stress, and contact pressure of assembly TKR. In addition, the colour ranges from blue to red, showing a minimum to a maximum value. This colour code serves to make the results visible to the user.

Table 5.1 Modifications made for Design 2 and Design 3

Design	Cross-Section	Description
Design 1	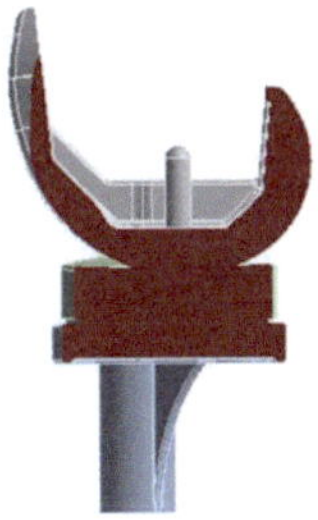	The maximum thickness of 11 mm and the minimum of 10 mm created a slight curve surface on the tibial insert, allowing a medium contact area with the femoral component
Design 2	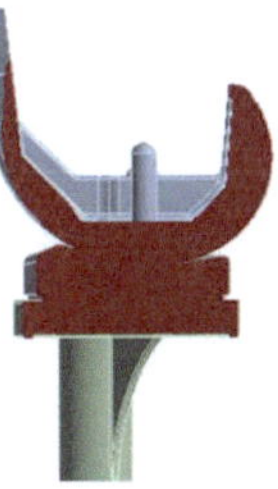	The maximum thickness of 11.2 mm and the minimum of 8.2 mm created an inclination surface on the tibial insert to offer a large contact area with the femoral component
Design 3	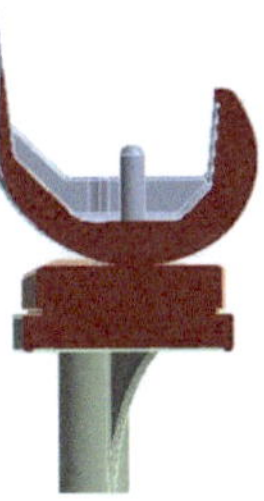	The flat surface with a thickness of 10 mm on the surface of the tibial insert gives a small contact area with the femoral component

5.5 Total Deformation (mm) of Modified Assembly TKR

Figure 5.8 shows the analysis result obtained for total deformation on the assembly TKR components at the angle flexion of 0°. According to the analysis, Design 1 measured the maximum deformation of 0.289 mm. Meanwhile, Design 2 has the maximum deformation of 0.207 mm, and Design 3 displayed the maximum deformation of 0.183 mm. These maximum deformations occurred at the upper part of the anterior femoral component for all the designs. The trend of maximum total deformations observed at the upper part of the femoral component corresponds with the study mentioned in the previous chapter. It was reasonable to justify that the deformations were acceptable for the simulation as the results were aligned closely with the trends.

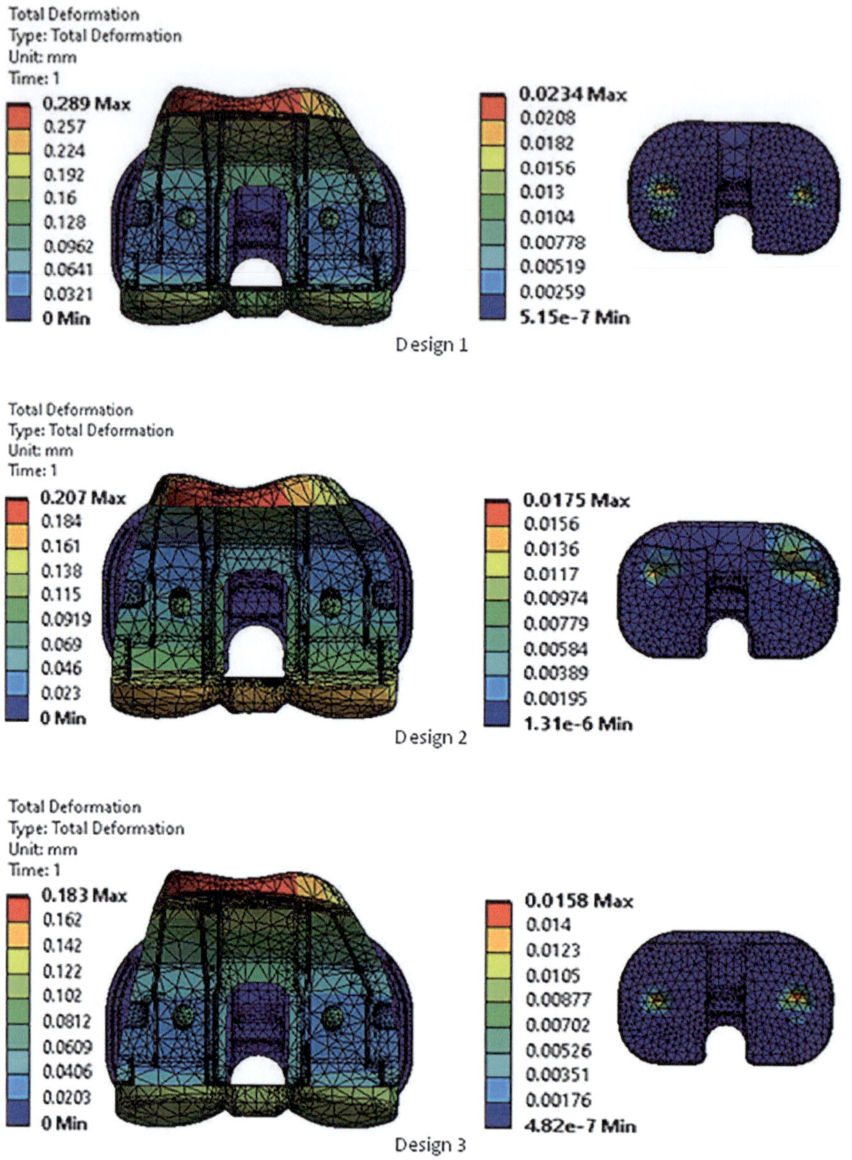

Fig. 5.8 Total deformation at the angle flexion of 0° for Design 1, Design 2, and Design 3

Figure 5.8 also shows the analysis of total deformation on the tibial insert at the angle flexion of 0°. The maximum deformation was centralised between the anterior and posterior of the tibial insert in Designs 1 and 3. Meanwhile, the maximum deformation observed in Design 2 was spread at the anterior medial of the tibial insert. It is due to the posterior roll movement of the femoral on the tibial insert observed during the simulation in Design 2. Not just that; due to the large contact area, the deformation was distributed wider in Design 2 compared to Designs 1 and 3. As a result, the femoral allows a wider flexion movement to the tibial. However, the anterior glide movement was observed in Designs 1 and 3 during the simulation. Consequently, this movement inhibits flexion of the femoral to the tibial.

Figure 5.9 shows the analysis result obtained for total deformation on the assembly TKR components at 90°.

According to Fig. 5.9, Design 1 measured the maximum deformation of 0.387 mm. Meanwhile, Design 2 has the maximum deformation of 0.223 mm, and Design 3 displayed the maximum deformation of 0.376 mm. These maximum deformations were observed at the upper part of the anterior femoral component in all the designs. As for observation, the posterior roll movement was shown during simulation in all the designs. Hence, the deformations at angle flexion of 90° have moved slightly to the posterior area of the tibial insert.

Moreover, Fig. 5.9 also shows the analysis of total deformation on the tibial insert at 90°. As for observation, the deformation in Design 2 was distributed towards the medial and lateral posterior of the tibial insert. Compared to Designs 1 and 3, which were centralised at the medial and lateral posterior of the tibial insert, it was observed that Design 2 has the maximum deformation at the post of the tibial insert. It might happen because the femoral tends to bend more at the inclination surface on the tibial insert. Consequently, the tendency moved the cam of the femoral to be in contact with the post of the tibial insert.

Figure 5.10 shows the analysis result obtained for total deformation on the assembly TKR components at the angle flexion of 135°. According to the display, Design 1 measured the maximum deformation of 0.923 mm. Meanwhile, Design 2 has the maximum deformation of 0.775 mm, and Design 3 displayed the maximum deformation of 0.789 mm. The maximum deformation still occurred at the upper part of the anterior femoral component. The posterior roll movement was observed at the angle flexion of 135° during the simulation, resulting in the femoral component moving towards the posterior section of the tibial insert.

Furthermore, Fig. 5.10 also shows the analysis of total deformation on the tibial insert at the angle flexion of 135°. As the flexion angle has increased to 135°, the deformations have occurred near the medial and lateral posterior of the tibial insert. Upon observation, the deformations were concentrated in a small area at the post of the tibial insert across all the designs. Note that it has appeared in Design 2 at the angle flexion of 90° only. In Design 2, the deformation was spread at the medial and lateral posterior sections despite not centralised at the posterior section like in Designs 1 and 3.

Figure 5.11 shows the result obtained for the analysis of total deformation on the assembly TKR components at the angle flexion of 165°. According to the analysis, Design 1 measured the maximum deformation of 1.018 mm. Meanwhile, Design 2 has the maximum deformation of 0.814 mm, and Design 3 displayed the maximum deformation of 1.144 mm. Upon observation, the maximum deformation still occurred at the upper part of the anterior femoral component at the angle flexion of 165°. It is because the posterior roll movement occurred to the femoral component during simulation in Design 2, resulting in the femoral component moving towards the posterior section of the tibial insert. However, the anterior glide movement was

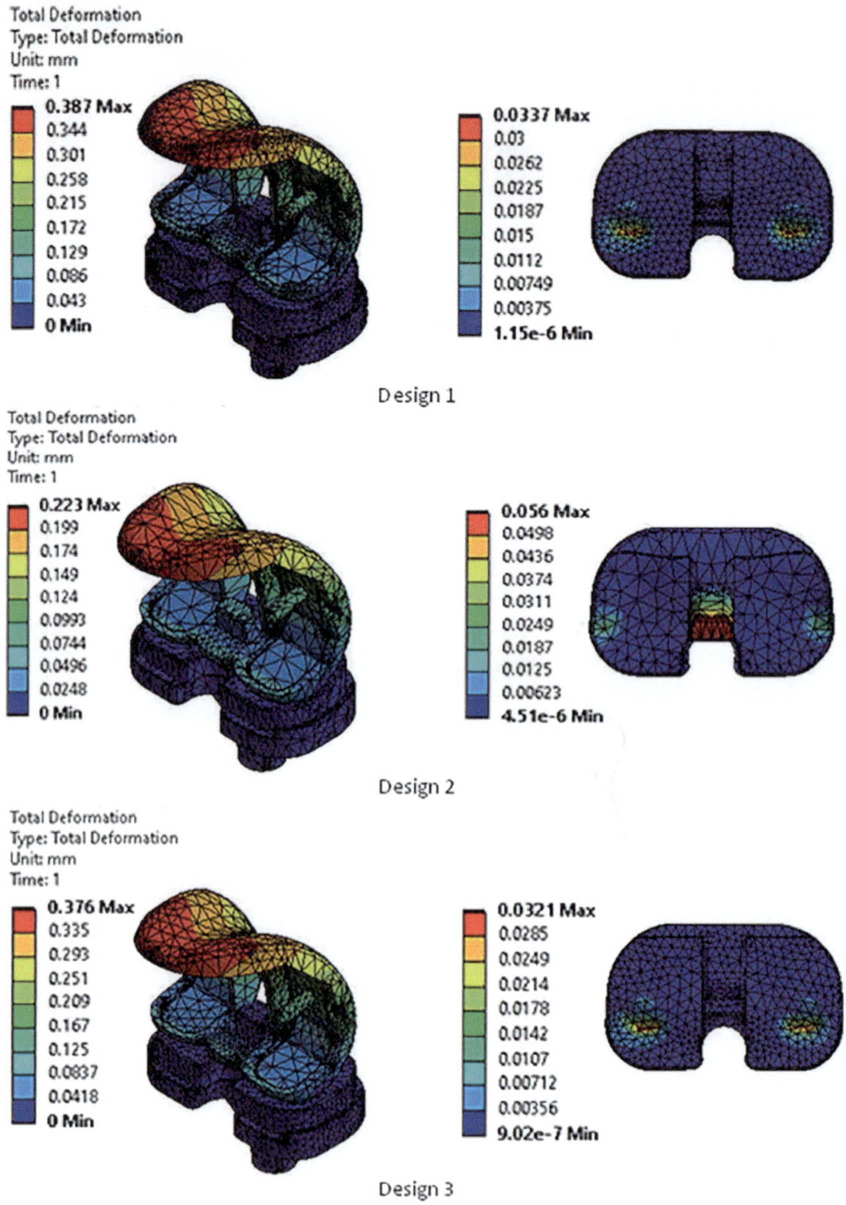

Fig. 5.9 Total deformation at the angle flexion of 90° for Design 1, Design 2, and Design 3

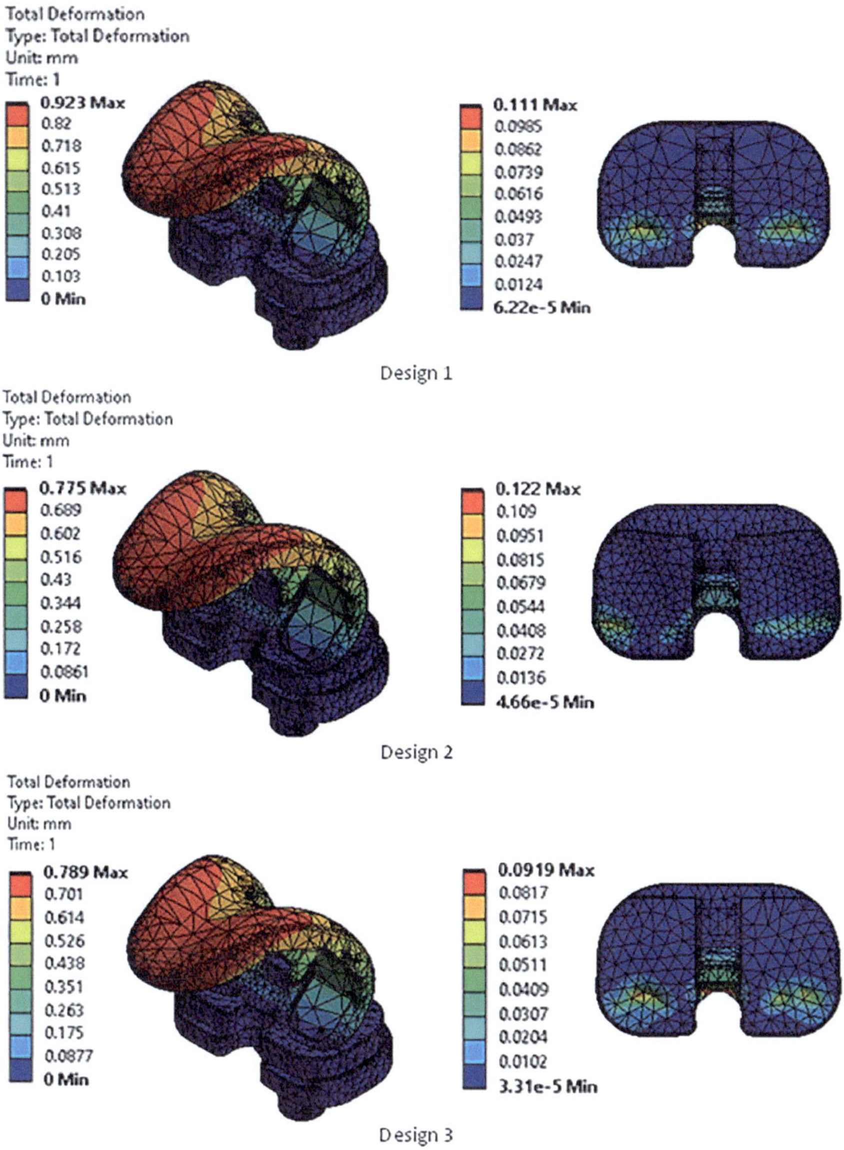

Fig. 5.10 Total deformation at the angle flexion of 135° for Design 1, Design 2, and Design 3

observed in Designs 1 and 3 during the simulation. Consequently, this movement inhibits flexion of the femoral to the tibial in Designs 1 and 3.

In addition, Fig. 5.11 also shows the analysis of total deformation on the tibial insert at the angle flexion of 165°. All the designs display deformations at the end of the medial and lateral posterior of the tibial insert. Note that as the angle flexion

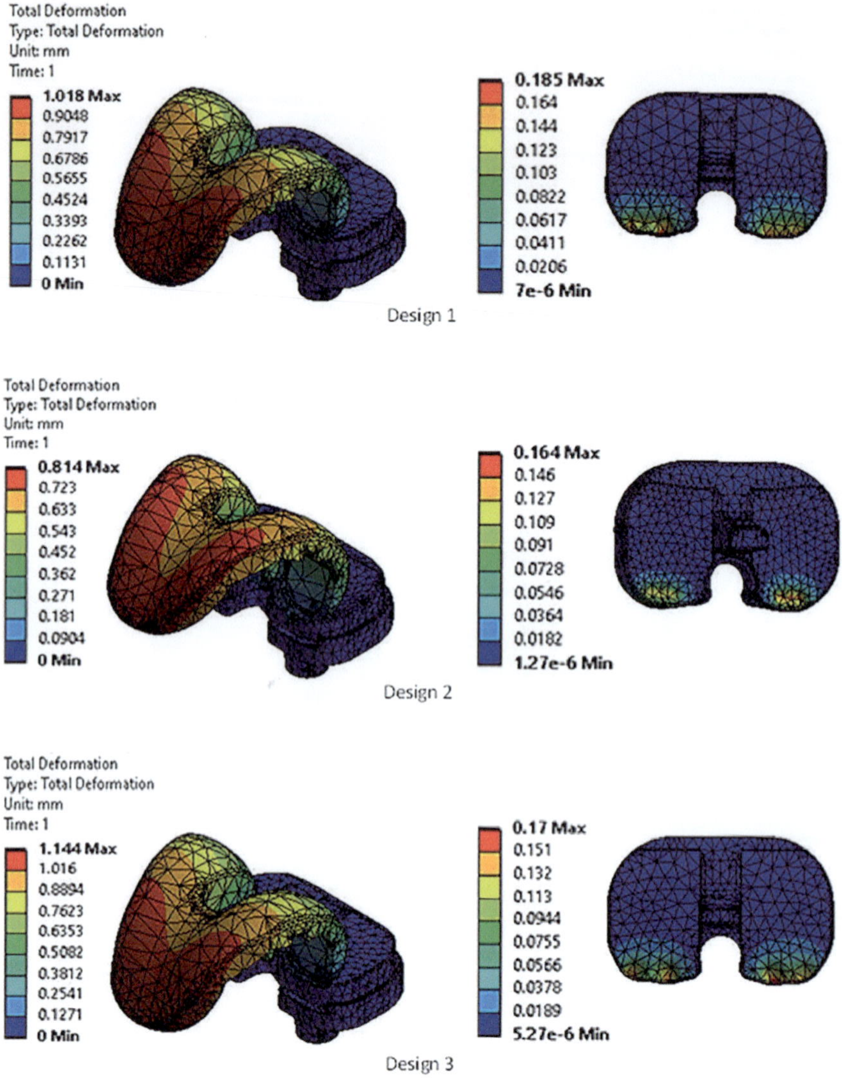

Fig. 5.11 Total deformation at the angle flexion of 165° for Design 1, Design 2, and Design 3

increases, the contact area becomes smaller. Therefore, increases in the deformation led to an increased failure rate of the tibial insert.

The outcome of total deformation for Design 1, Design 2, and Design 3 was compared in the graph, as shown in Fig. 5.12.

Upon observation, Design 2 has the lowest deformation at the angle flexion of 90°, 135°, and 165°. However, at the angle flexion of 0°, the anterior glided movement was observed in Design 1 during the simulation, which lessens the deformation in Design

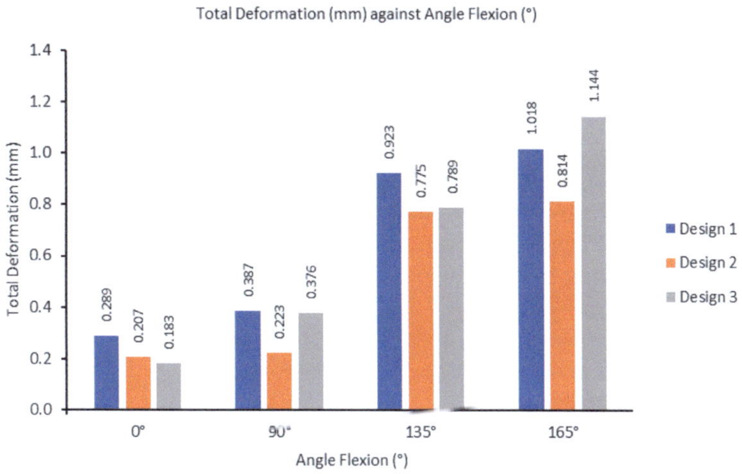

Fig. 5.12 Graph of Total Deformation (mm) Against Angle Flexion (°)

2 by 28.32%. However, this movement allows Design 3 to reduce the deformation by a 13.18% difference from Design 2. Note that the trend of total deformation increased against the angle flexion, which stated that total deformation inclined against the increasing angle flexion on different conformity of the tibial insert.

Despite that, the strong point of Design 2 is introducing the inclination surface on the tibial insert to provide a larger contact area between the femoral component and the tibial insert. Besides that, the posterior roll movement observed in Design 2 at the angle flexion of 0°, 90°, 135°, and 165° during simulation increases the ability of flexion to the modified TKR design.

To conclude, Design 2 has the smallest deformation at the angle flexion of 90°, 135°, and 165° compared to Designs 1 and 3. The modification made in Design 2 to achieve high flexion with the lowest deformation was achieved. Hence, it can be concluded that the large deformation occurred at the small contact area.

5.6 Von Mises Stress (MPa) of Modified Assembly TKR

Figure 5.13 shows the analysis result obtained for von Mises stress on the assembly TKR components at the angle flexion of 0°. The maximum von Mises stress measured in Design 1 is 6.418 MPa, and the minimum is 3.42E-06 MPa. Meanwhile, Design 2 has the maximum von Mises stress of 7.054 MPa and a minimum of 4.24E-06. Lastly, Design 3 measured the maximum von Mises stress of 4.618 MPa and the minimum of 3.72E-06 MPa. The maximum von Mises stresses were more distributed at the medial and lateral of the femoral condyles. However, in Design 2, the maximum von

Mises stress was scattered at the medial anterior of the femoral condyle. Therefore, the distribution trend of von Mises stress at the femoral condyles corresponds to the study.

According to Fig. 5.13, the von Mises stresses were centralised at the posterior of the tibial insert in Designs 1 and 3. Meanwhile, the von Mises stress was found to be distributed at the medial anterior of the tibial insert in Design 2. It is due to the posterior roll movement of the femoral on the tibial insert observed during the simulation in Design 2. However, the anterior glide movement was observed in Designs 1 and 3 during the simulation. As a result, this movement inhibits femoral flexion to the tibial.

Figure 5.14 shows the analysis result obtained for von Mises stress on the assembly of TKR components at the angle flexion of 90°.

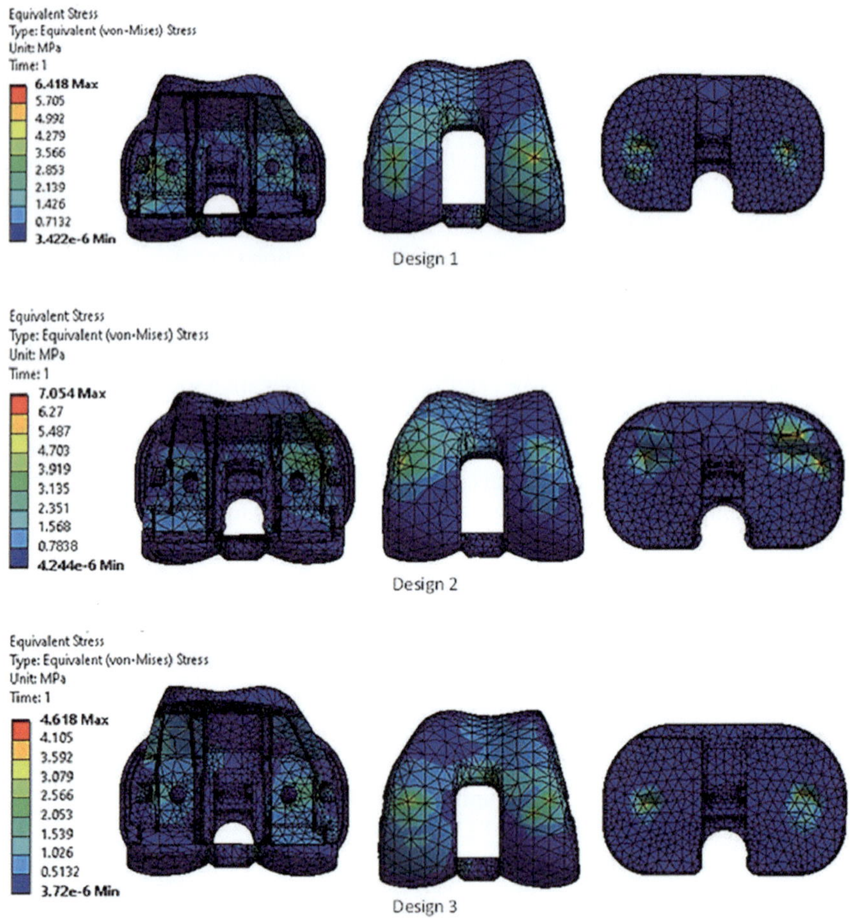

Fig. 5.13 Von Mises Stress at the angle flexion of 0° for Design 1, Design 2, and Design 3

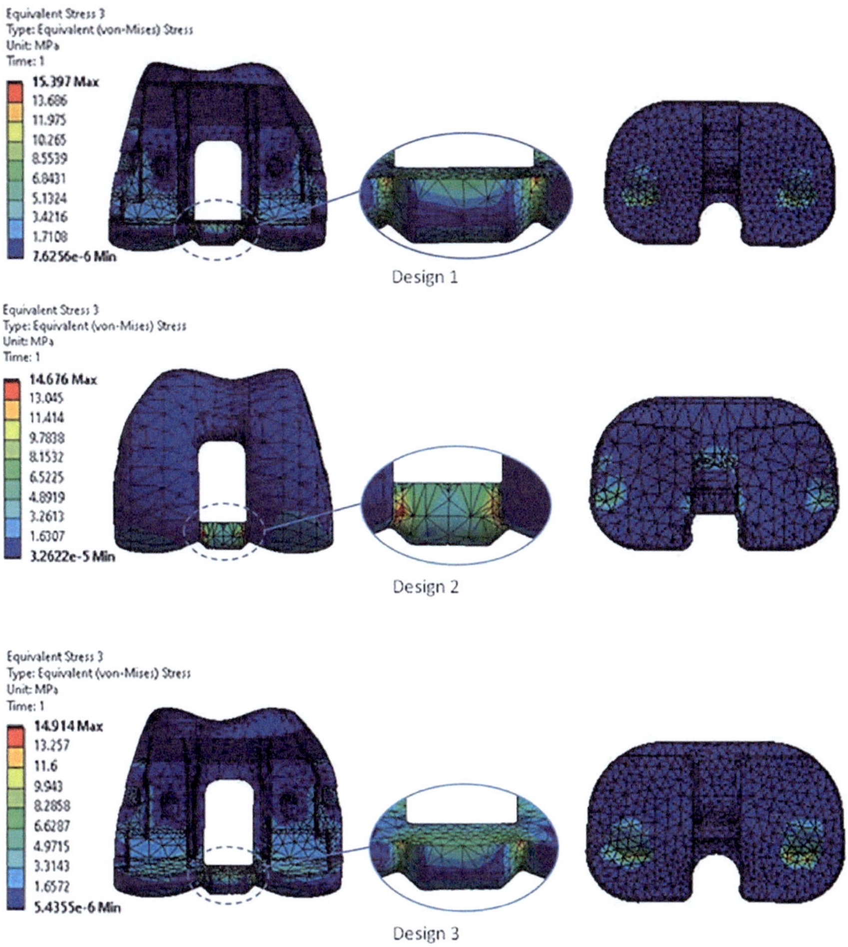

Fig. 5.14 Von Mises Stress at the angle flexion of 90° for Design 1, Design 2, and Design 3

According to Fig. 5.14, the maximum von Mises stress measured in Design 1 is 15.397 MPa, and the minimum is 4.83E-06 MPa. Meanwhile, Design 2 experienced the maximum von Mises stress of 14.676 MPa and a minimum of 3.26E-05. Lastly, Design 3 showed the maximum von Mises stress of 14.914 MPa and a minimum of 5.44E-06 MPa. In all the designs, the maximum von Mises stresses were confined near the cam at the upper part of the posterior femoral component.

Furthermore, the von Mises stresses were concentrated at the posterior of the tibial insert in Designs 1 and 3. Meanwhile, in Design 2, the von Mises stress was spread at the medial and lateral posterior of the tibial insert. Upon observation, all the designs showed posterior roll movement during simulation. Hence, the von Mises stress at the angle flexion of 90° has moved slightly to the posterior area of the tibial insert. Also, it was observed that Design 2 has the maximum deformation at the post of the tibial insert. It might happen because the femoral tends to bend more at the inclination surface on the tibial insert. Consequently, the tendency moved the cam of the femoral to be in contact with the post of the tibial insert.

Figure 5.15 shows the analysis result obtained for von Mises stress on the assembly TKR components at the angle flexion of 135°. The maximum von Mises stress measured in Design 1 is 89.518 MPa, and the minimum is 3.70E-06 MPa. Meanwhile, Design 2 has the maximum von Mises stress of 82.274 MPa and a minimum of 4.22E-06. Lastly, Design 3 measured the maximum von Mises stress of 81.719 MPa and the minimum of 4.89E-06 MPa. It was observed that the von Mises stresses distributed more at the cam of the posterior femoral component.

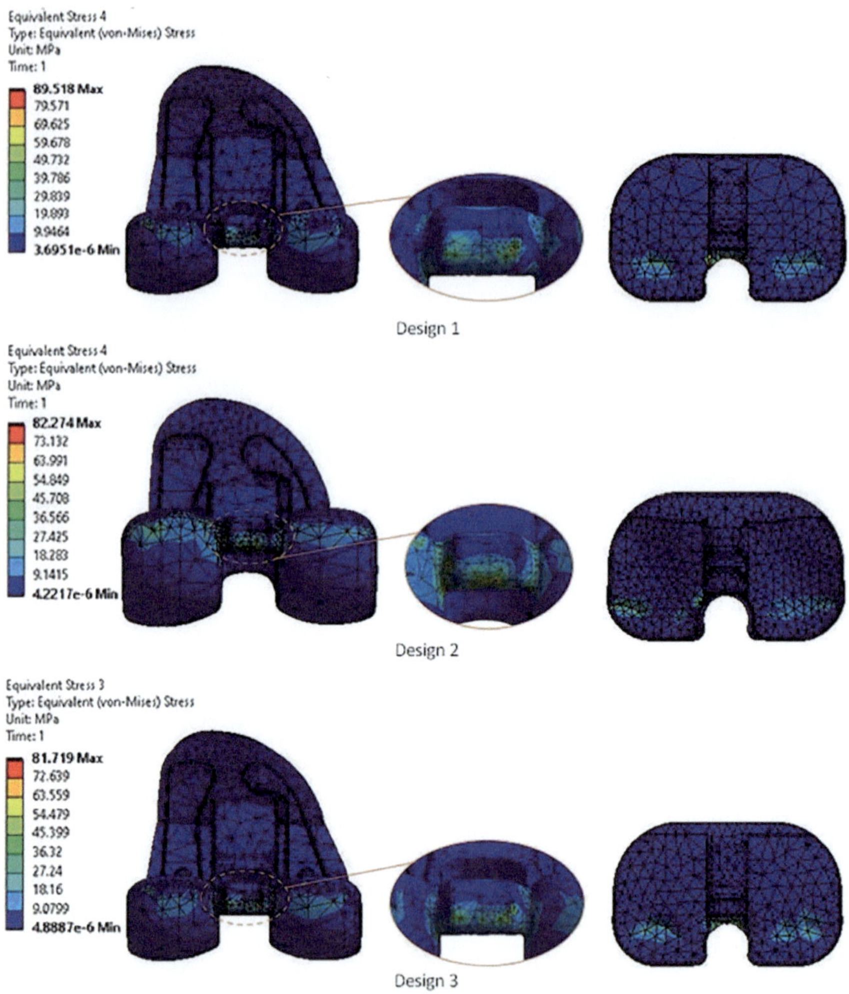

Fig. 5.15 Von Mises Stress at the angle flexion of 135° for Design 1, Design 2, and Design 3

According to Fig. 5.15, in Designs 1 and 3, the von Mises stresses accumulated at the tibial insert's posterior. Meanwhile, in Design 2, the von Mises stress was found to be distributed towards the medial and lateral posterior of the tibial insert. Upon observation, the von Mises stresses were found at the post of the tibial insert in a small area in all the designs. It is due to the posterior roll movement being observed at the angle flexion of 135° during simulation, resulting in the femoral component moving towards the tibial insert cam.

Figure 5.16 shows the analysis result obtained for von Mises stress on the assembly TKR components at the angle flexion of 165°. The maximum von Mises stress measured in Design 1 is 124.355 MPa, and the minimum is 2.980E-06 MPa. Meanwhile, Design 2 experienced the maximum von Mises stress of 104.161 MPa and a minimum of 3.18E-06. Lastly, Design 3 measured the maximum von Mises stress of 104.823 MPa and the minimum of 3.10E-06 MPa.

According to Fig. 5.16, the von Mises stresses were confined at the upper part and around the cam of the posterior femoral component in all the designs. Design 2 exhibits posterior roll movement during the simulation at the angle flexion of 165°, allowing more femoral flexion on the tibial insert. In Designs 1 and 3, the von Mises stresses were accumulated at the posterior of the tibial insert. Upon observation, the anterior glide movement was observed during the simulation in Designs 1 and 3. Consequently, this movement restrains flexion of the femoral to the tibial component.

All the designs display von Mises stress at the end of the medial and lateral posterior of the tibial insert. Note that as the angle flexion increases, the contact area becomes smaller. Therefore, increases in the von Mises stress influence the risk of fracture for the tibial insert.

The outcome of von Mises stress for Design 1, Design 2, and Design 3 was compared in the graph, as shown in Fig. 5.17.

According to the graph, Design 2 has the smallest von Mises stress at the flexion angle of 90° and 165°. Meanwhile, Design 3 had the smallest von Mises stress at the flexion angle of 0° and 135°. Moreover, Design 2 has more significant von Mises stress than Design 3, with a 52.74% difference at 0° and a 0.68% difference at 135°. However, at 90°, Design 2 has a smaller von Mises stress with a 1.60% difference and 0.63% difference at 165° compared to Design 3. Note that the trend of von Mises stress increased against the angle flexion, as mentioned in Sect. 2.5.3, which stated that von Mises stress inclined with the increasing of angle on different conformity of the tibial insert.

Despite that, the maximum von Mises stresses on the femoral in all the designs at the angle flexion of 0°, 90°, 135°, and 165° were lower than the defined yield strength of 900 MPa for titanium alloy, as mentioned in Sect. 2.4. Therefore, the application of titanium alloy for femoral and tibial components is excellent in TKR for preventing the risk of fracture. However, the maximum von Mises stress on the tibial insert used polyamide 12 to apply the material. Table 5.2 shows the maximum von Mises stress on the tibial insert in all the designs at the angle flexion of 0°, 90°, 135° and 165°.

Regarding Table 5.2, the maximum von Mises stresses on the tibial in all the designs at the angle flexion of 0°, 90°, 135°, and 165° were lower than the defined yield strength of 55 MPa for polyamide 12, as mentioned in Sect. 2.4. However, at the angle flexion of 135° and 165°, the maximum von Mises stress was still lower than

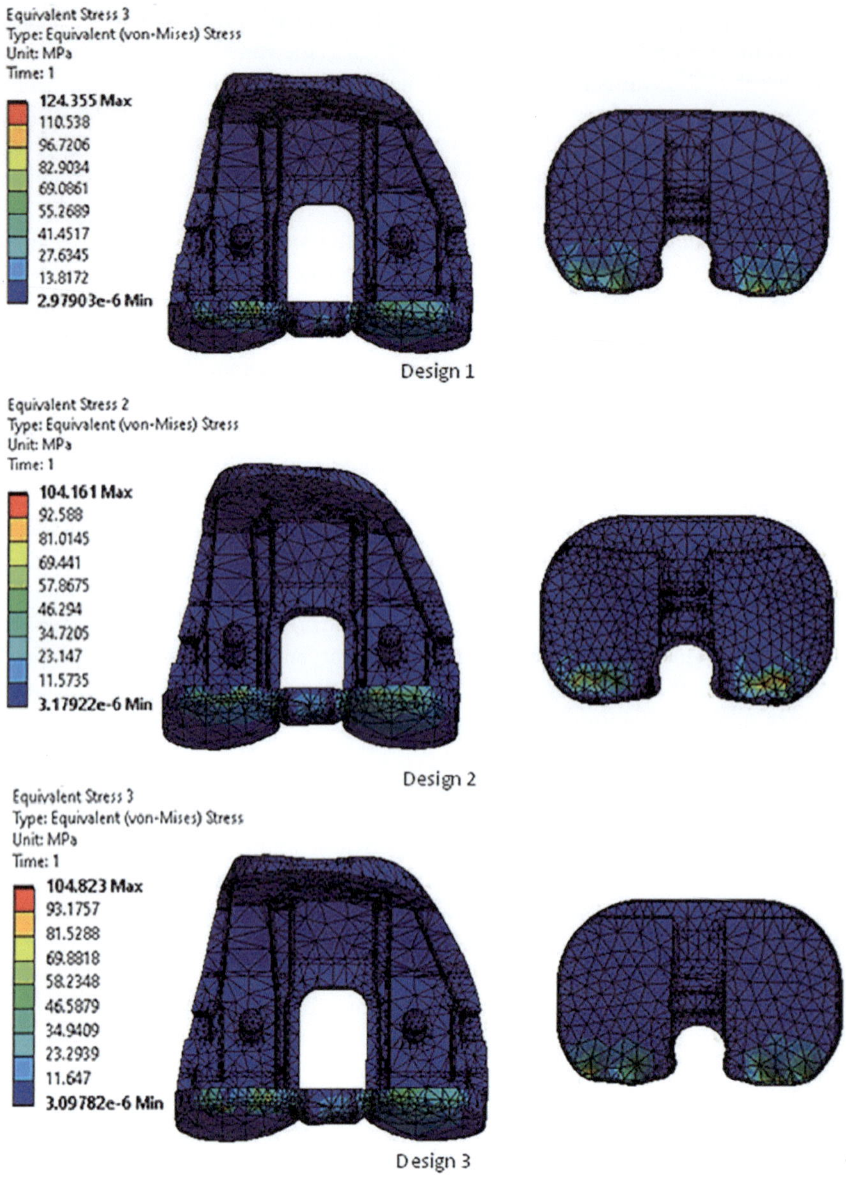

Fig. 5.16 Von Mises Stress at the angle flexion of 165° for Design 1, Design 2, and Design 3

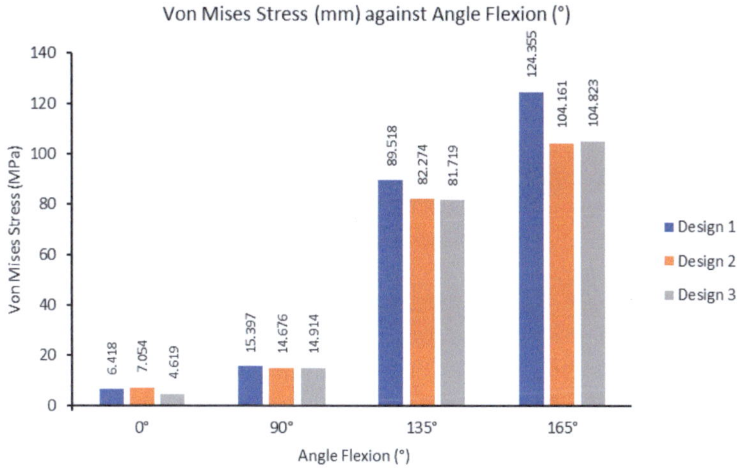

Fig. 5.17 Graph of von Mises Stress against Angle Flexion

Table 5.2 Von Mises Stress (MPa) on the Tibial Insert

Angle (°)	Design 1 (MPa)	Design 2 (MPa)	Design 3 (MPa)
0°	6.418	4.636	4.618
90°	11.927	10.927	8.175
135°	47.813	53.186	47.020
165°	47.615	36.346	46.311

the defined yield strength in a small amount, such as in Design 2 at 135°. Applying polyamide 12 for the tibial insert in all the designs is acceptable in TKR to prevent fracture risk.

To conclude, Design 2 has the lowest von Mises stress at angle flexion of 90° and 165° compared to Designs 1 and 3. It also prevents the risk of fracture in the tibial insert when performing high flexion at the angle flexion of 165°. The modification made in Design 2 to achieve high flexion with the lowest von Mises stress was achieved at the angle flexion of 0°, 90°, and 165°. Hence, it can be concluded that the maximum von Mises stress should be lower than the yield strength of the material used to prevent the risk of material fracture.

5.7 Contact Pressure (MPa) of Modified Assembly TKR

Figure 5.18 shows the analysis result of contact pressure on the assembly TKR components at the angle flexion of 0°. According to Fig. 5.18, Design 1 experienced the maximum contact pressure of 17.993 MPa. Design 2 measured the maximum contact pressure of 13.473 MPa, while Design 3 had the maximum pressure of 15.129 MPa. Upon observation, the contact pressure in Design 2 experienced the maximum pressure at the medial anterior of the tibial insert. It is due to the posterior roll movement of the femoral on the tibial insert observed during the simulation in Design 2.

In Design 1, the maximum contact pressure was concentrated at the medial and lateral posterior of the tibial insert. Meanwhile, in Design 3, the maximum contact pressure was concentrated at the lateral posterior of the tibial insert. Nevertheless, the anterior glide movement was observed in Designs 1 and 3 during the simulation. Consequently, this movement inhibits flexion of the femoral to the tibial. To conclude, the distribution trend of contact pressure at the medial and lateral of the tibial insert corresponds to the study.

Figure 5.19 shows the analysis result of the contact pressure on the assembly TKR components at the angle flexion of 90°.

According to Fig. 5.19, Design 1 experienced the maximum contact pressure of 34.276 MPa. Design 2 measured the maximum contact pressure of 28.5298 MPa. Meanwhile, Design 3 has a maximum contact pressure of 19.293 MPa. In Design 2, the contact pressure was distributed at the medial and lateral of the femoral component and tibial insert. It is due to the posterior roll movement of the femoral on the tibial insert observed during the simulation in Design 2. However, in Designs 1 and 3, the contact pressures were concentrated at the medial and posterior of the femoral component. It is due to the anterior glide movement observed in Designs 1 and 3 during the simulation. As a result, this movement inhibits femoral flexion to the tibial.

Figure 5.20 shows the analysis result obtained from contact pressure on the assembly TKR components at the angle flexion of 135°.

According to Fig. 5.20, Design 1 experienced a maximum contact pressure of 199.8604 MPa. Design 2 measured the maximum contact pressure of 124.133 MPa, while Design 3 had a maximum of 163.282 MPa. The contact pressure between the femoral component and the tibial insert was distributed near the upper part of the posterior femoral component. Upon observation, the contact pressures were found at the post of the tibial insert in a small area in all the designs. It is due to the posterior roll movement being observed at the angle flexion of 135° during simulation, resulting in the femoral component moving towards the tibial insert cam.

Figure 5.21 shows the result obtained for the analysis of contact pressure on the assembly TKR components at the angle flexion of 165°.

According to Fig. 5.21, Design 1 experienced the maximum contact pressure of 201.093 MPa during the analysis. Design 2 measured the maximum contact pressure of 195.156 MPa, while Design 3 had the maximum of 239.963 MPa. These maximum contact pressures between the femoral component and the tibial insert were observed at the end of the upper part of the posterior femoral component in all the designs.

Upon observation, Design 2 exhibits the posterior roll movement during the simulation at the angle flexion of 165°, allowing more femoral flexion on the tibial insert. However, the anterior glide movement was observed during the Designs 1 and 3 simulations. Consequently, this movement restrains flexion of the femoral to the tibial component. Therefore, as the angle flexion increases, the contact area becomes smaller, which increases the contact pressure between the femoral and tibial insert.

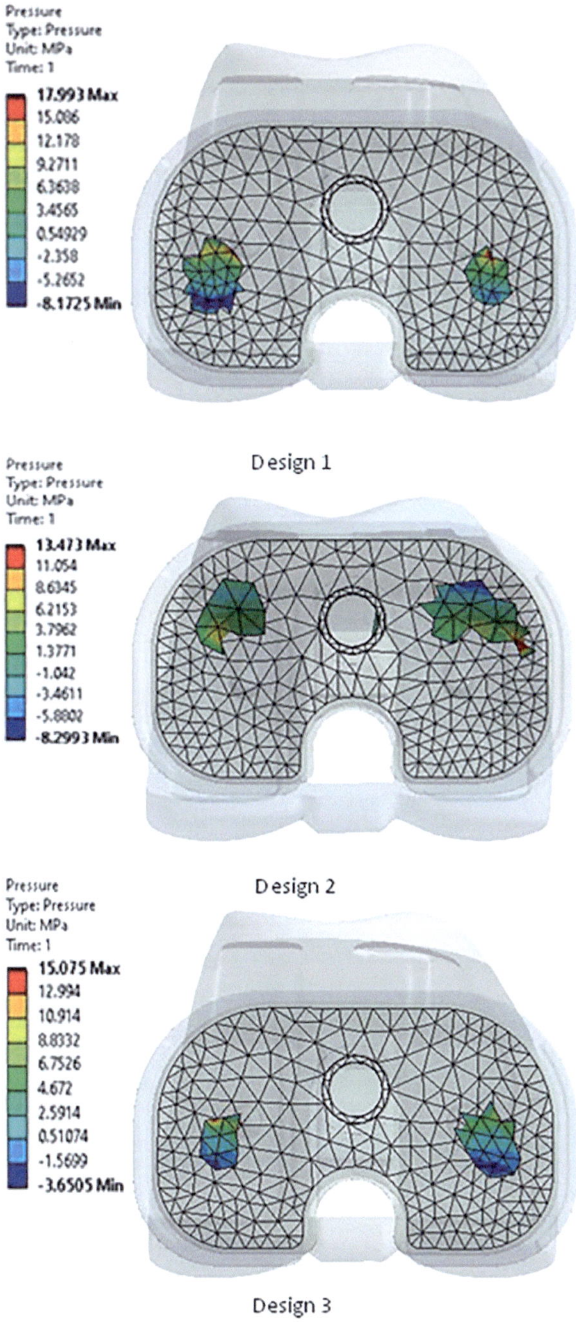

Fig. 5.18 Contact pressure at the angle flexion of 0° for Design 1, Design 2, and Design 3

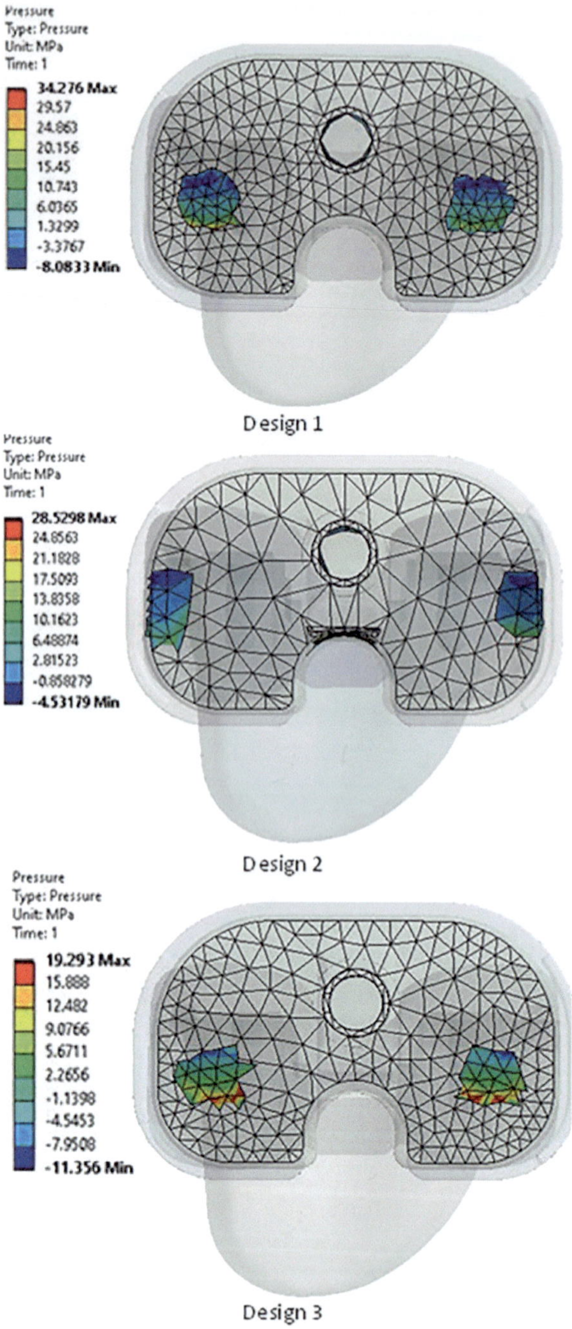

Fig. 5.19 Contact pressure at the angle flexion of 90° for Design 1, Design 2, and Design 3

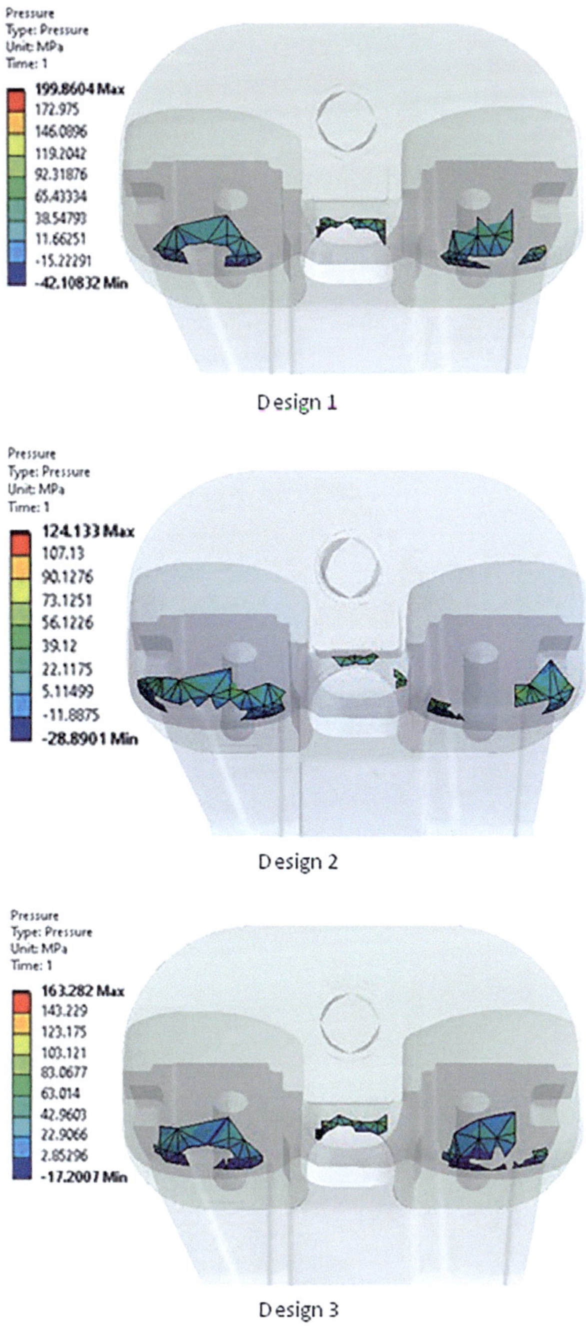

Fig. 5.20 Contact Pressure at the angle flexion of 135° for Design 1, Design 2, and Design 3

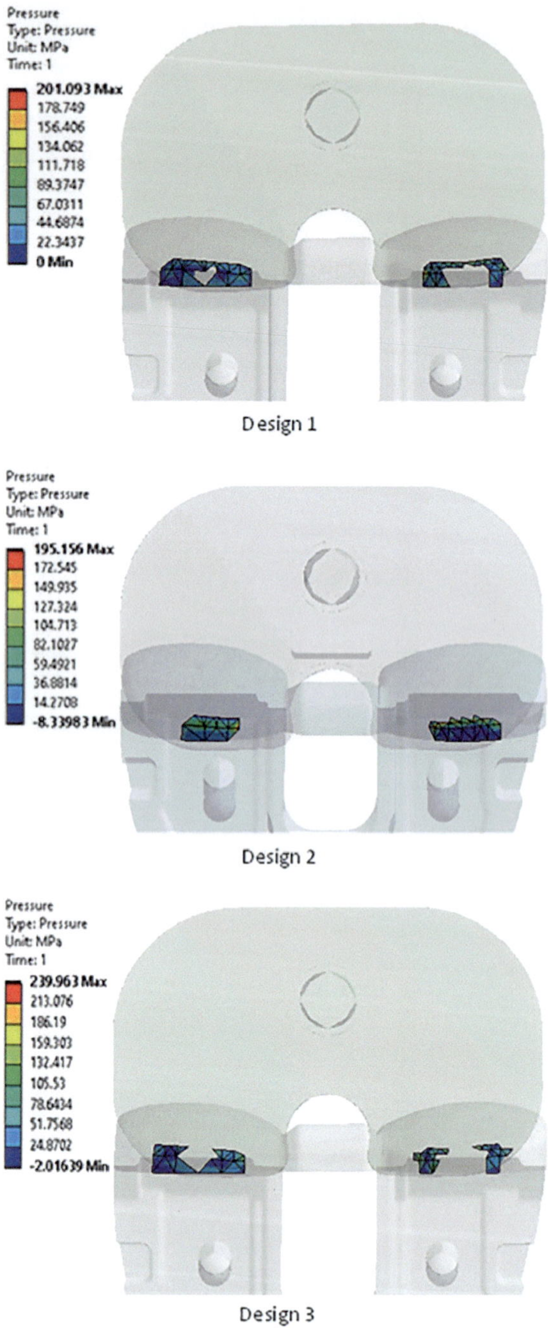

Fig. 5.21 Contact Pressure at the angle flexion of 165° for Design 1, Design 2, and Design 3

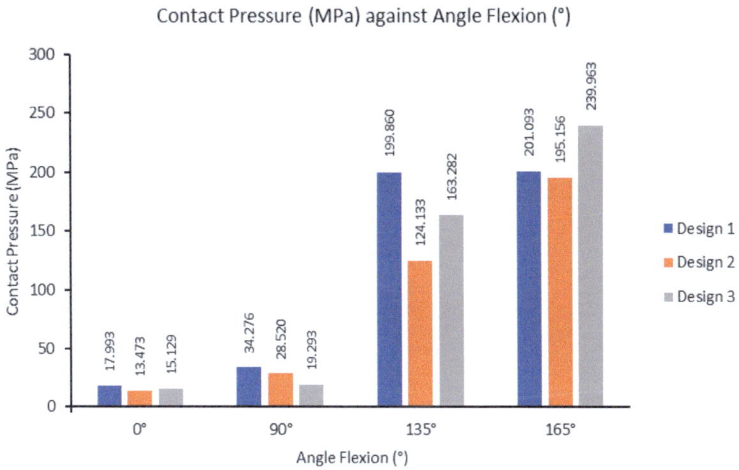

Fig. 5.22 Graph of contact pressure against Angle Flexion

The outcome of contact pressure for Design 1, Design 2, and Design 3 was compared in the graph, as shown in Fig. 5.22.

Upon observation, Design 2 has the lowest contact pressure compared to Designs 1 and 3, except at the angle flexion of 90°. The inclination surface on the tibial insert in Design 2 provided a smaller area than in Designs 1 and 3 during the flexion angle of 90°. Not just that; it can be observed that Design 2 has a larger contact pressure with a 47.88% difference when compared to Design 3, and a smaller contact pressure with a 16.76% difference than Design 1 at 90°. Note that the trend of contact pressure increased against the angle flexion, regarding contact pressure on different conformity of the tibial insert.

To conclude, Design 2 has the lowest contact pressure at 0°, 135°, and 165° compared to Designs 1 and 3. The modification made in Design 2 to increase high flexion with the lowest contact pressure was achieved. Hence, it can be concluded that the small contact area influences the higher contact pressure.

5.8 Comparison Results Between the Compression Testing and the Finite Element Analysis for Total Deformation

Results obtained from the FEA for Designs 1, 2, and 3 were compared to find the best knee prosthesis that could increase the range of flexion. The comparison shows that Design 2 has less deformation, von Mises stress, and contact pressure at the angle flexion of 165° compared to Designs 1 and 3. The significance of lessening deformation, von Mises stress, and contact pressure at the angle of 165° shows that the knee prosthesis does not have difficulties flexing beyond certain angles, which

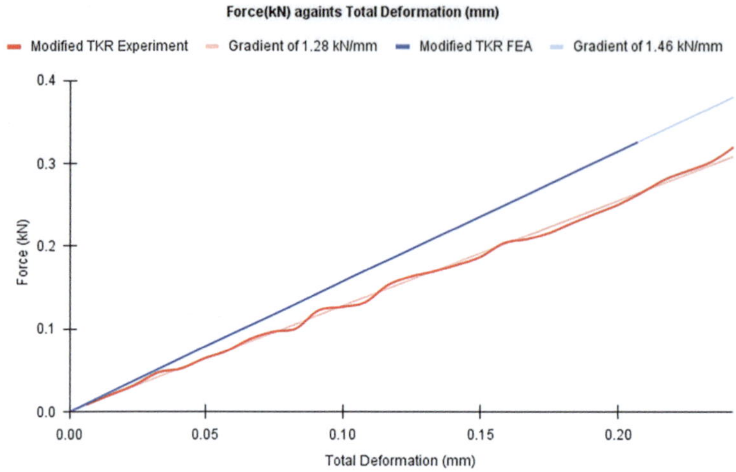

Fig. 5.23 Graph of force against Total Deformation

could prevent materials failure. Therefore, Design 2 was selected and named as the modified TKR design and printed using AM technologies.

The modified TKR design is compared with the experimental compression results to verify the design and validate the results obtained from the FEA. The jigs for femoral and tibial components are fabricated to fit in the testing machine to secure the printed modified TKR. The reason is to obtain the assembly TKR in a static mode with the angle flexion of 0° to represent standing. In this case, the deformation is measured under the force of 326.18 MPa at the angle flexion of 0° to ensure the validity of the comparison. Figure 5.23 shows the graph comparison of experimental results on the compression test and results of FEA for modified TKR design.

According to Fig. 5.23, it can be observed that the FEA results have a linear line because the force implemented on the TKR components measured an outcome for the force applied to the TKR components. Meanwhile, the experimental result shows irregular fluctuations as it ascends. The actuator's force displacement was used at 0.5 mm/min to load on the femoral. The irregular fluctuations because of the multiple results were measured until the load reached 326.18 N.

According to Fig. 5.23 and Table 5.3, the maximum deformation of the modified TKR experiment is 0.242 mm. Meanwhile, modified TKR FEA has the maximum deformation of 0.207 mm at the angle flexion of 0°, as stated in Sect. 4.3.1. The maximum deformation on the experimental testing was consistent with the result of FEA, with a percentage error of 14.46%. The decline in accuracy of the FEA compared to the prior study is due to the design of jigs or fixtures to hold the femoral and tibial components. It was assumed that the femoral jig holding the femoral component should remain static. However, the placement of the femoral jig to the femoral component could have been moving during the experiment because of the compressive loading, which led to lower accuracy results.

Table 5.3 Tabulate results of maximum deformation of modified TKR experiment and FEA

Angle (°)	Force (N)	Modified TKR experiment (mm)	Modified TKR FEA (mm)
0°	326.18	0.242	0.207

Furthermore, the materials and manufacturing processes can significantly impact the strength of the printed TKR. The prior study used commercial knee implants with the standard material properties during compression testing. In this study, the layer thickness, printing temperature, and post-processing treatments of the AM processes may affect the strength of fabricated TKR, which can influence the final material properties. Hence, the material properties implemented during the FEA could also be affected, leading to lower accuracy between the FEA and experimental results.

Furthermore, the gradients of the graphs represent the stiffness of TKR. The FEA result shows 1.28 kN/mm stiffness or 1280 kN/m, while the experimental result shows 1.46 kN/mm stiffness or 1460 kN/m. The simulation and experimental stiffness of TKR show a percentage difference of 14.06%, which may indicate fair agreement. In this situation, the compressive load of materials can bear the loads before fracture, which also decreases the risk of fracture for the modified TKR design. In final thoughts, the percentage difference between total deformation and stiffness of TKR shows the same tendency in terms of displacement and von Mises stress.

5.9 Summary

To summarise the result and discussion section, the selection design process was introduced to determine the best type of TKR for increasing the range of motion by using a Pugh chart. The process of designing for modifications of TKR components was thoroughly explained, focusing on the contact between the femoral and tibial insert. For instance, Design 1, Design 2, and Design 3 were labelled to represent small, medium, and large contact areas. The FEA results in terms of total deformation, von Mises stress, and contact pressure were elaborated thoroughly with four different angles of flexion, such as 0°, 90°, 135°, and 165°, to represent standing, kneeling, squatting, and sitting on feet positions.

The FEA results between unmodified and modified TKR were compared comprehensively. Design 2, selected as the final modified TKR design, went through fabrication using additive manufacturing processes because of the capabilities to produce complex geometry and save time. The comparison of deformation between compression testing and FEA was compared and explained. The results show a 14.46% difference between both deformations with stiffness differences of 14.06%. As the results have been discussed here, the next chapter will conclude the research objectives with additional future work recommendations.

References

1. Z. Biomet, The Personalized Knee Story. Zimmer Biomet (2001). [Online]. Available: https://www.zimmerbiomet.com/content/dam/zimmer-biomet/medical-professionals/knee/persona-knee-system/0890.1-GLBL-en%20Persona%20The%20Personalized%20Knee%20Story-v16-72dpi.pdf. Accessed 12 April 2022
2. S.V.D. Groes, Aseptic Loosening and Anterior Knee Pain in High-Flexion Total Knee Arthroplasty. Gildeprint—The Netherland (2014)

Chapter 6
Conclusion and Recommendation

This chapter summarises the purpose of the study with the main final thoughts and recommendations for future works.

6.1 Final Thoughts

The research of developing a design for high flexion and finite element analysis in total knee replacement was proposed with three objectives for finding the best design and analysis for high flexion in TKR. The first objective was achieved by developing a design of the TKR component using CAD software for a better range of motion. Developing a design for TKR involved design selection and design modification, which were explained in the procedure section in detail. The femoral component was designed with the post-cam mechanism, and the tibial component was developed with the tube for inserting the stem of the tibial insert for the locking mechanism. Meanwhile, the modified tibial insert has a maximum thickness of 11.2 mm and a minimum of 8.2 mm, creating an inclination surface on the tibial insert to provide a large contact area between the femoral component and the tibial insert.

The second objective was performing finite element analysis on the assembly of TKR components using ANSYS software. Several modified designs underwent meshing to obtain outcomes in total deformation, von Mises stress, and contact pressure. Static structural analysis was conducted on the modified TKR components with four different flexion angles to represent standing, kneeling, squatting, and sitting on feet. The result of FEA shows that the inclination surface on the tibial insert lessens the deformation by 28.32%, 42.29%, 16.03%, and 20.05%, at the angle flexion of 0°, 90°, 135°, and 165° respectively, compared to the unmodified tibial insert. The von Mises stress was decreased by 4.68%, 8.09%, and 16.24% at the angle flexion

I. M. Arrif et al., *High-Flexion Knee Prosthesis*,
SpringerBriefs in Applied Sciences and Technology,
https://doi.org/10.1007/978-981-96-9878-3_6

of 90°, 135°, and 165°, respectively, compared to the unmodified tibial insert. Meanwhile, the contact pressure was reduced by 25.12%, 16.79%, 37.89%, and 2.95% at the angle flexion of 0°, 90°, 135°, and 165° respectively, compared to the unmodified tibial insert. The final modified design shows that the tibial insert's inclination surface was selected for fabrication using additive manufacturing processes based on the FEA results.

The last objective was to perform the compression test on the fabricated TKR components. Additive manufacturing was utilised to produce the TKR components using selective laser sintering to print titanium alloy material for femoral and tibial components and fused deposition modelling to print polyamide material for the tibial insert. Compression testing was carried out to compare the results for both deformations. The load applied during compression testing was 326.18 N, representing a standing position at 0°. The maximum deformation for compression testing was compared to the final modified design of FEA for total deformation at the angle flexion of 0° with a percentage difference of 14.46%. The result between the simulation and experimental shows good agreement, considering the consistency of the FEA results for all designs in terms of total deformation.

The results of FEA presented were in the linear elastic region and below the yield stress of titanium alloy and polyamide, which are 930 MPa and 55 MPa, respectively. It also indicates no fracture risk and failure for all the modified designs of TKR components during analysis. Furthermore, a large contact area introduced to modify TKR components was achieved successfully, as the outcome shows a minor deformation and contact pressure during analysis. Hence, the modified TKR components achieved better results compared to the reference design, the NexGen LPS-Flex Knee, by increasing the thickness of the tibial insert to achieve a flexion angle of 165°. In final thoughts, the modification of the high-flexion TKR design was completed by having the results of FEA reduced in total deformation, von Mises stress, and contact pressure but also having consistency in stiffness and deformation between simulation and experiment.

6.2 Recommendations

This research study has taken into consideration many aspects and made a significant effort to achieve the research objectives. Nevertheless, there is potential for improvement and recommendations for future studies related to the design and development of high flexion for TKR using the FEA approach. Further improvements and recommendations include increasing the thickness of the posterior femoral condyle to enhance the range of motion, as this would lengthen the contact area between the posterior femoral and tibial insert. Besides that, the femur and tibia bones during FEA to mimic the natural knee joint better and implement rigid body dynamics in simulation to measure the movement of knee joints and implants precisely. Secondly, the jig for the compression test can be further improved to allow it to perform tests at various angle flexions of TKR and the suitability of machine use for TKR testing could measure stress. Lastly, AM processes could be involved in the study by producing TKR components rather than outsourcing to enhance the working procedure.

Index

I. M. Arrif et al., *High-Flexion Knee Prosthesis*, SpringerBriefs in Applied Sciences and Technology, https://doi.org/10.1007/978-981-96-9878-3

MIX
Papier aus verantwortungsvollen Quellen
Paper from responsible sources
FSC® C105338

If you have any concerns about our products,
you can contact us on
ProductSafety@springernature.com

In case Publisher is established outside the EU,
the EU authorized representative is:
Springer Nature Customer Service Center GmbH
Europaplatz 3, 69115 Heidelberg, Germany

Printed by Libri Plureos GmbH
in Hamburg, Germany